Understanding Broadband over Power Line

Understanding Broadband over Power Line

Gilbert Held

Auerbach Publications
Taylor & Francis Group
Boca Raton   New York

Auerbach Publications is an imprint of the
Taylor & Francis Group, an informa business

Published in 2006 by
Auerbach Publications
Taylor & Francis Group
6000 Broken Sound Parkway NW, Suite 300
Boca Raton, FL 33487-2742

International Standard Book Number-10: 0-8493-9846-0 (Hardcover)
International Standard Book Number-13: 978-0-8493-9846-9 (Hardcover)
Library of Congress Card Number 2005058919

Library of Congress Cataloging-in-Publication Data

Held, Gilbert, 1943-
    Understanding broadband over power line / Gilbert Held.
        p. cm.
    Includes bibliographical references and index.
    ISBN 0-8493-9846-0 (alk. paper)
    1. X10 (Power line control protocol) I. Title.

TK5103.4.H45 2006
621.382--dc22   2005058919

Taylor & Francis Group
is the Academic Division of Informa plc.

Visit the Taylor & Francis Web site at
http://www.taylorandfrancis.com

and the Auerbach Publications Web site at
http://www.auerbach-publications.com

Dedication

Over the past two decades I have had the privilege to teach a series of graduate courses focused on communications technology for Georgia College and State University. Teaching graduate school has enabled me to both convey information as well as learn from the inquisitive minds of students.

Long ago, when I commenced my first full-time job at IBM Corporation, many desks were most notable by the placement of a sign that simply stated the word, "Think." For more than 20 years, my graduate school students have made me remember the information that sign at IBM conveyed. In recognition of their inquisitive nature, this book is dedicated to the students of Georgia College and State University.

Contents

Acknowledgments

A long time ago, after I completed my first manuscript, I became aware of the many persons involved in the book production process. From the typing of an initial manuscript to the editing process and galley page production, through the cover design process and binding effort, there is a literal army of men and women whose efforts are crucial in producing the book you are now reading. I would be remiss if I did not acknowledge their efforts.

First and foremost, every book idea will come to naught unless an author works with an editor who has the foresight and vision to back an effort focused on an emerging technology. Once again, I am indebted to Rich O'Hanley at CRC Press for supporting my writing efforts.

The preparation of a manuscript is a long and lengthy process. Although this author has many laptop and notebook computers, long ago he gave up modern technology for pen and paper. Regardless of where one travels, the differences in electrical outlets and airline policies do not have an effect on pen and paper. Although this method of manuscript generation may appear awkward in today's era of electronic gadgets, as long as I have paper and pen I do not have to worry about whether or not I can use my computer on an airline or if my outlet converter will mate to the receptacle used in a hotel. Of course, my penmanship leaves a lot to be desired and makes me truly grateful for the efforts of my wife, soul mate, and excellent typist, Beverly. Commencing her effort on a 128-kB Macintosh to type my first book almost 30 years ago, Beverly now uses Microsoft Word under Windows XP on both desktop and notebook computers to not only type this author's manuscripts, but also type the index for each book.

Once a manuscript reaches the publisher, a number of behind the scenes efforts occur. Once again, I am indebted to Claire Miller for guiding the manuscript through the book production process. Concerning that process, I would also like to thank the CRC Press team in Boca Raton, Florida, for their efforts in reviewing and editing my manuscript as well as for guiding the galley pages into the book you are reading.

Preface

Just when you thought high-speed Internet access was limited to cable television and the local telephone company, you now have an additional option to consider. Broadband over power lines (BPL) represents an emerging technology that enables electric utilities to provide support for high-speed data communications over their infrastructure into our homes and offices. To paraphrase Barry Goldwater, we now have a choice instead of an echo.

As I will note in this book, BPL represents a technology that dates to the late 1970s and early 1980s, when several vendors developed products to enable electric wiring in homes and offices to be used to share what were then expensive printers and plotters. Although the first generation of BPL products developed for home and office use never achieved a degree of success, advances in microelectronics and filter design resulted in the development of two similar technologies that enable high-speed data communications in both the home and office over existing electrical wiring and over power lines, which make up the infrastructure of electric utilities. In this book I will examine both technologies, because the two are closely interrelated and the ability to transmit data over power lines is not particularly useful if a connection method is not available to enable home and business subscribers to easily make an Internet connection through the utility infrastructure. Concerning the connection method, at the time this book was researched and written, there were two methods being used in field trials. The most popular method was to use the electrical wiring in the home or office as a mechanism to communicate to and from the power line bringing electricity into the facility. A second method involved the use of a wireless LAN (local area network) access point connected to the electric utility infrastructure. By placing the access point at a centralized location within a cluster of homes or offices, the utility can provide a communications capability to many subscribers, allowing each subscriber to use relatively low-cost IEEE 802.11-compliant client wireless LAN cards to connect to the access point.

I will make only one assumption as we investigate the operation of BPL communications. That assumption is that my readers have a variety of backgrounds, ranging in scope from network engineers to life science majors as well as other persons who may not have a scientific background. Thus, to make this book as practical as possible for an audience with a diverse background, I have written several chapters to provide tutorial information. For example, in this book we will examine the manner by which electric circuits and wireless LANs can be used to provide access to power lines operated by electric utilities. In addition, to ensure all readers have a common level of knowledge concerning the electric utility infrastructure and home wiring, we will review power line operations. In doing so, we will examine how electricity is generated and delivered into our homes and offices as well as how the wiring inside our structures forms individual circuits that are routed to a common circuit breaker or fuse box. An appreciation for how several technologies operate will allow for a greater appreciation for the manner by which BPL technology has the potential to represent an emerging ubiquitous communications technology that, when deployed, can provide individual consumers, network managers, and LAN administrators with another solution to satisfy their high-speed networking requirements.

As a professional author, I highly value reader feedback. You can contact me either through my publishers, whose address is on the jacket of this book, or directly via e-mail at gil_held@yahoo.com. Let me know if I dwelt too long on a particular topic or if I did not devote enough information concerning a particular area, or just provide me with your general comments concerning this book. Because I frequently travel, I may not immediately answer your letter or e-mail; however, I will answer all persons within a week or two.

Gilbert Held
Macon, GA

Chapter 1

Understanding Broadband over Power Lines

Similar to the first chapter in any technical-oriented book, the purpose of this chapter is to acquaint the reader with the topic of the book. In this chapter we will turn our attention to obtaining an appreciation for how data communications can occur over power lines installed to transport electricity. In addition, we will examine the potential use of broadband over power lines (BPL) as a mechanism for providing Internet access to both homes and offices. Because there are several methods that can be used to obtain Internet access, we will also review the competition to broadband communications over power line technology. This information will then be used as a mechanism to better understand the rationale for obtaining another method of high-speed communications, which is the focus of this book. Because new technology cannot be expected to be problem free, we will also discuss problems associated with the transmission of data over circuits originally intended to convey electricity.

1.1 Overview

Broadband over power lines, which is the title of this book, represents an emerging technology that can provide high-speed Internet access to the home or office through the use of an electrical outlet. Referred to as BPL, broadband over power lines theoretically has the ability to enable data to be transmitted over power lines into homes and offices at data rates between 500 kbps and 3 Mbps, which is equivalent to most Digital Subscriber Line (DSL) and cable modem transmission rates. Thus, BPL provides an emerging alternative to conventional methods of obtaining high-speed Internet access.

The key reason for the excitement concerning BPL technology is the fact that virtually every home and office is connected to a power grid and contains electrical wiring. Thus, any mechanism that provides the potential to transmit high-speed data over existing electrical wiring has the potential to provide a truly ubiquitous method to access the Internet. That said, let's turn our attention to how this technology evolved.

Evolution

Although most readers may think of BPL as a relatively recent technology, in actuality it dates back approximately 25 years to the development of the personal computer. At that time local area networks (LANs) were still in their infancy and the most common method used to share what were then costly printers and plotters was through the use of parallel and serial mechanical switches.

Mechanical Switches

The use of serial and parallel interface switches enabled two or three computers to be connected to a printer or plotter, although obviously only one computer could access the printer or plotter at any point in time. In addition, manual intervention was required to change the setting on a switch to enable a different computer to access the shared plotter or printer.

Electric Wiring

Recognizing the need for an automatic method to share peripheral devices in the home or small office over extended distances, several vendors introduced communications products during the late 1970s and early 1980s that used existing electric wiring as a transport mechanism. Such devices consisted of a plug that was connected to the serial port of a computer and inserted into an electrical outlet. The oversized plug contained a digital-to-radio frequency modulator and demodulator. Through the use of radio frequency (RF) communications, digital data transmitted via the computer's serial port was modulated and transmitted over the power line to another plug-type device. That device was connected to a printer or plotter in the same or a different room, in effect providing a home networking capability over existing electrical wiring.

The initial series of products developed during the late 1970s and early 1980s to enable data transmission over home power lines never achieved any significant degree of success. The reasons varied by product, but can probably be summed up in two areas: a lack of miniaturization, which resulted in rather bulky adapters, and the relatively slow data rate provided by the use of a connection to the serial port of a computer.

The HomePlug Standard

The low data rate reflected limits on the serial port of computers during the 1970s and 1980s, which limited the transfer rate to, at best, 19,200 bps during the 1970s and 115,000 bps during the 1980s, with the higher transfer rate resulting from the introduction of buffered universal asynchronous receiver-transmitters (UARTs).

Since the introduction of BPL adapters, advances in microelectronics and the development of the USB port have resulted in a renewed interest in home networks over the electrical circuits in the home and small office. USB ports provide a data transfer capability ranging from ten to a hundred times greater than the most capable UART. Concerning advances in microelectronics, instead of the bulky adapters used during the 1970s and 1980s, more modern adapters for use in the home and small office are relatively small and simple to insert into a standard electrical outlet. The combination of microelectronics and the ability to connect to the higher speed USB port now standard on desktop and laptop computers resulted in several vendors forming the Home-Plug Powerline Alliance, which released its specification for high-speed power line networking products that enable Ethernet-class operations over standard home electrical wiring.

Broadband Over Power Lines

Although the HomePlug standard represents an important step for the future of transmission over electrical circuits, home networking represents a different technology from BPL where the existing power grid infrastructure is used to provide high-speed broadband Internet access to homes and businesses. As we will note later in this book, transformers are used to raise the voltage on power lines routed from power generation plants. As high voltages are transmitted over long distances, transformers are also located where branch lines carrying lower voltages are routed to geographic locations containing clusters of homes and offices. At those locations, additional transformers are employed to reduce voltage to 120 volts, which is then routed into homes and offices.

Due to high-voltage lines and transformer coupling presenting a different environment from the home or office, a different technology emerged to solve the problem associated with moving data over power lines. Although many principles associated with transmitting data over power lines outside the home and electrical circuits within the home are similar, equipment used on power lines differs from equipment used within the home. One of the most obvious differences concerns the fabrication of equipment to withstand the elements. When designed for outdoor use, equipment requires shielding from the elements to include fabrication that makes the devices waterproof. When used indoors, similar performing equipment does not require the ability to withstand rain, snow, fog, and other weather conditions. Another obvious difference concerns the transport mechanism. In a home or office, data will be modulated to flow over a specific type of electrical wiring. In comparison, when data is modulated to flow over power lines, the type of modulation used will vary based on the transport facility. As we will note later in this chapter as well as later in this book, electric utilities have installed tens of thousands of miles of optical fiber along their high-voltage power lines. Originally used exclusively for power line monitoring and internal communications, those optical fiber facilities can also be used to provide a data transport facility for customers by increasing the capacity of the fiber through the use of wavelength division multiplexing (WDM) and dense wavelength division multiplexing (DWDM). Because optical fiber is usually not available on electrical branches where medium- and low-voltage lines are distributed toward residential and commercial customers, electric utilities will convert optical transmission into electrical transmission via RF modulation over their power lines routed to homes and offices. Thus, BPL technology represents multiple modulation methods, whereas internal home or office use of electrical wiring represents a single modulation method.

During the 1990s, several European utilities conducted trials involving the transmission of data over power lines. Although the initial trials produced mixed results, advances in technology resulted in additional trials occurring during the turn of the new century. In the United States the momentum associated with transmission of data over utility power lines significantly increased as the use of the Internet increased. Unfortunately, many large utility operators expanded their operations into the so-called merchant energy field during the years of the so-called Western energy crisis. When the energy crisis abated, many utility operators were left owing billions of dollars for power plants that were not economical to operate, which resulted in an electrical utility financial crisis. Between 2000 and 2004 several power plant operators went belly-up, declaring bankruptcy, and large utility operators who expanded their power plant portfolios had significant liquidity problems, forcing them to curtail their investments in noncore operations. Fortunately, as the recession of 2001 receded and demand for power increased, utilities in the United States and other locations once again examined the use of their infrastructure for the transmission of data. As we will note later in this book, utility operators initiated a number of field trials as the economy rebounded, and some organizations now offer Internet access in competition with cable and telephone companies.

As we build on our knowledge as we move from one chapter to the next in this book, we will become aware of the similarities and differences between the home networking and power line environments. That said, let's turn our attention to the basic technology that enables data to be transmitted over power lines.

1.2 Fundamental Concepts

The ability to transmit data over power lines in many ways is based on the concept by which telephone companies observed that the standard wire pair routed into homes and offices could be used to transmit data at rates up to and beyond 1 Mbps. The telephone companies recognized the fact that the twisted-pair telephone line routed into homes and offices was capable of supporting a frequency range up to approximately 1 MHz. Because a telephone conversation uses only approximately 3 kHz of bandwidth, it becomes possible to transmit data by modulation occurring at frequencies beyond those used to convey voice. This technique, with which voice is transported at one set of frequencies while data is transported by modulation

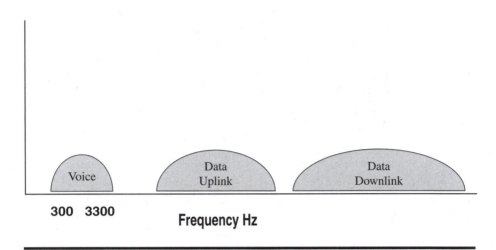

Figure 1.1   Frequency utilization of a telephone company local loop line.

occurring at a different set of frequencies, is referred to as frequency division multiplexing (FDM) and serves as the foundation for the development of DSL and cable modem technologies. When we discuss BPL modulation methods in some detail in Chapter 3, we will note the similarity of the technology to DSL technology with respect to the use of FDM to separate data transmission from the primary use of the line, which is for voice when a DSL is used and for power when a BPL system is employed. For now, let's turn our attention to how DSL technology operates.

Digital Subscriber Lines

During the late 1980s, telephone companies' suppliers introduced a series of DSL products. Each product took advantage of the fact that the telephone company local loop used only the frequencies between 300 and 3300 Hz for a voice conversation. By splitting the remaining frequency into two bands, one for the uplink and the other for the downlink, it became possible to transmit high-speed data over a standard local loop even while a telephone conversation was taking place.

Figure 1.1 illustrates how the frequency of a telephone company local loop is employed when DSL transmission occurs on the line. Note that the larger frequency band (referred to as the downlink band) is used to support data transmission from the telephone company to the subscriber. Because most DSL usage is for Internet access, vendors realized that Web pages would flow downstream toward the subscriber. In comparison, because relatively short page requests in the form of URLs flow toward the Internet, the uplink band uses less frequency than the downlink band. Because the data rate is proportional to available bandwidth, the downlink band supports a higher data transfer rate than the uplink frequency band. At the subscriber's home or office, a DSL modem is used to modulate and demodulate data transmitted over the two frequency bands. Because the two bands are not symmetrical, this technology is referred to as an Asymmetrical Digital Subscriber Line (ADSL). Note that the three frequency bands shown in Figure 1.1 represent an FDM system occurring over a twisted-wire circuit routed from a telephone company office to a subscriber.

Power Line Operation

Using the telephone line as a familiar point of reference, let's turn our attention to power line operations and the manner by which data can be transmitted concurrent with electricity. Standard alternating current (AC) is transmitted at a frequency of 60 Hz in North America and at 50 Hz in Europe and many other locations throughout the world. This means that, similar to a telephone company local loop, electrical lines have almost all of their frequency available for utilization for other purposes, to include data transmission. Consider Figure 1.2, which illustrates the frequency use of a power line. Assuming the power line

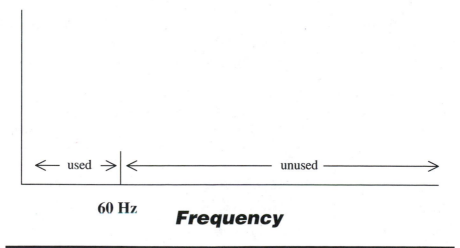

Figure 1.2 Frequency use of a power line.

is used in North America, frequencies beyond 60 Hz are unused. Thus, the evolution of data transmission over the unused frequencies of power lines is based on the same concepts that enable DSL to provide millions of subscribers with high-speed Internet access.

Overcoming High Voltages

Unlike the telephone local loop, which has a limited transmission distance, power lines allow for data transmission over extended distances. However, this results in the degradation of analog signals used to modulate data. This problem can be further exacerbated when power lines transport high voltages. To overcome the problem associated with high-voltage lines, many electric utilities are using fiber-optic cables running in parallel with those lines. Originally, the fiber-optic lines were installed for power monitoring and control purposes. Although most utilities use only a small fraction of the capacity of their fiber, it is relatively simple to expand the capacity of existing fiber by many orders of magnitude through the use of WDM equipment. Because the data is transmitted in the form of light pulses, it is not disturbed by the high voltage of the power lines. At distribution locations the optical signals are converted to electrical signals, which are then carried over medium-voltage and low-voltage lines.

Medium- and Low-Voltage Line Operation

As data flows over medium- and low-voltage metallic conductors, transmission distance becomes an issue. In addition, unlike telephone company lines that are shielded, power lines are not. Because data is modulated for transmission using analog technology over medium- and low-voltage lines, at periodic distances the analog signals must be amplified. This amplification process is similar to the manner by which telephone companies use amplifiers to boost analog signals. However, instead of being called amplifiers, BPL terminology refers to such devices as repeaters. In Chapter 3 we will examine some of the technical terms associated with BPL technology.

The Home and Office Connection

Once the analog signal arrives within the vicinity of its ultimate destination, there are two methods being used in trials to provide a communications connection into homes and offices. Those methods are wired and wireless transmission. When the wired option is used, residential and business customers must obtain a power line modem that, when plugged into an electrical outlet, modulates and demodulates data flowing over the power line routed into the building. As we will note later in this book, the modulation method used on low-voltage lines provides an access method that is used to support transmission from many users over medium-voltage lines. When a wireless connection is employed, the electric utility typically installs an IEEE 802.11b or g access point on a utility pole at a location where it can serve multiple homes or offices. Then, each user in a home or office that subscribes to the service uses a wireless network adapter in a desktop or notebook computer to connect to the access point. This type of infrastructure bypasses the need for a modulation method on low-voltage lines as well as allows a node on the medium-voltage line to serve many customers via over-the-air transmission, referred to as WiFi or wireless fidelity.

Modulation Methods

Although the IEEE is working on a standard that will cover the lower two layers of the protocol suite used for transmission of data over power lines, that standard will probably be promulgated in a year or two. In the interim, field trials performed in North America and Europe have used several modulation methods selected by developers as being responsive to the conditions associated with transmitting data over low- and medium-voltage power lines. The two primary modulation methods selected for many field trials are code division multiple access (CDMA) and orthogonal frequency division multiplexing (OFDM). Although we will review modulation methods in some detail in Chapter 3, a few words are now in order concerning the difference between the two.

CDMA

Code division multiple access (CDMA) represents a modulation method employed with cellular phones. Originally, mobile telephone technology was developed using analog techniques, and early systems were referred to as analog mobile phone systems, or AMPS for short. Due to the need to support a growing base of mobile phone users, equipment vendors developed technologies that enabled the spectrum licensed for mobile telephone operations to be used more efficiently. One of the technologies developed was CDMA.

OFDM

A second modulation method used in field trials of systems developed to transmit data over low- and medium-voltage power lines is orthogonal frequency division multiplexing (OFDM). The evolution of OFDM can be traced to what at one time was high-speed 9600 bps modem technology, developed by a vendor named Telebit Corporation. Telebit introduced the first 9600 bps modem designed for use over the public switched telephone network (PSTN) during the 1980s. Since then, OFDM has evolved and is primarily used in IEEE 802.11a and 802.11g wireless LANs to provide a high-speed data communications capability in both the 2.4-GHz and 5-GHz frequency bands.

When comparing CDMA and OFDM, one must consider several tradeoffs, which we will focus on in Chapter 3. For the moment, we can note that CDMA can provide a higher data transmission capability than OFDM. However, because OFDM uses multiple carriers orthogonal to one another, it has more resilience to noise than CDMA. In Chapter 3 we examine each modulation method.

Current Problems

The development of a BPL transmission facility requires utilities to use their power lines to transmit data over those facilities. Although the use of fiber-optic lines routed along the main utility high-voltage transmission route does not present an interference problem due to the use of optical transmission, the same cannot be said for the use of analog modulation over medium- and low-voltage metallic facilities. In such situations a utility faces a minimum of two problems. First, the cost of bypassing transformers and adding equalizers to overcome attenuation problems needs to be considered. Second, because power lines are not shielded, they act as antennas when data is modulated and transmitted over those facilities. This results in the power line functioning as a radiating antenna, which results in interference that can render certain types of radio systems unusable if they are located in close proximity to a power line. In Chapter 2 we will examine the structure and routing of utility power lines to include the use of transformers, and in Chapter 3 we will discuss how power lines emit electromagnetic radiation. In Chapter 4 we will discuss a relatively recent FCC (Federal Communications Commission) ruling that can be expected to promote the use of BPL technology as a competitor to DSL, cable modem, and other high-speed networking technologies. Concerning those other networking technologies, in the next and concluding section of this chapter we will review competitive Internet access technologies.

1.3 Competitive Internet Access Technologies

In this concluding section of this chapter, we will turn our attention to both existing and emerging Internet access technologies. In doing so, we will compare each technology to the use of BPL, because the latter is the focus of this book. However, prior to discussing competitive Internet access technologies, a few words are in order concerning BPL. Once we obtain an appreciation for the benefits of BPL technology, then we can easily compare and contrast it with other Internet access technologies.

Advantages of BPL

The use of BPL offers several advantages. First, because the technology enables existing power lines outside the home to function as a data transmission medium, the utility operator can use its existing infrastructure. Although amplifiers are required as well as equipment to bypass transformers, converting power lines so that they can carry data does not require the degree of an upgrade that cable TV infrastructure upgrades required to support cable modem technology. Inside the home, data can be carried over the existing electrical wiring, which means that every room in the house has the capability to access the Internet. Thus, the use of inside and outside electrical wiring for data transmission provides the ability for homeowners or small businesses to use the electrical wiring within their facilities without modification. In fact, it becomes possible for a utility that signs up a customer to use its BPL technology to simply mail the customer a BPL modem that can be plugged into an outlet within the home or office. Thus, the customer does not have to wait for the service provider to install equipment nor does the customer have to modify their existing wiring because the electrical codes followed in all modern locations require several outlets in each room. Now that we have an appreciation for

Table 1.1 Competitive Internet Access Technologies

PSTN
Cable modem
DSL
Satellite
WiMax

some of the advantages associated with BPL technology, let's turn our attention to competitive Internet access technology. As we do so, we will compare and contrast each technology to the use of BPL technology.

Table 1.1 lists five Internet access technologies that can be considered as being competitors to BPL. The first Internet access technology, PSTN, has a limited data rate in comparison to the other competitive technologies. However, because access to the PSTN is available from just about every home and office, it provides a foundation for discussing other competitive technologies.

PSTN

The use of the public switched telephone network (PSTN) to obtain a communications capability is as ubiquitous as homes and offices that have electrical service. Although V.92 modems designed for use over the PSTN are stated as having a 56-kbps data transfer capability, in reality the maximum data transfer rate one obtains is typically between 41 and 44 kbps.

The key advantage associated with the use of the PSTN is its availability. A subscriber to an online PSTN Internet access plan can obtain a connection from any location that has a PSTN connection. Thus, it becomes possible to access the Internet from the home and office as well as when traveling, with a single Internet account. Concerning the cost of PSTN Internet access accounts, they are the most economical of all Internet access technologies. Although AOL, Microsoft, and other premium Internet access provider costs are in the low $20s per month range, several nationwide vendors now offer unlimited Internet access via the PSTN at monthly rates under $10. Thus, Internet access via the PSTN, although providing the lowest data transfer capability among the competitive methods listed in Table 1.1, is also the lowest cost method.

Cable Modem

Currently high-speed Internet access is dominated by the use of cable modem technology. During the late 1990s through the turn of the millennium, cable operators spent tens of billions of dollars upgrading their infrastructure to support two-way amplification of signals, which enabled the support of cable modem technology. Although cable modem service is now offered by just about every medium- and large-scale cable operator in North America, it is not ubiquitous. In fact, cable service is available to only 60–65 percent of homes in North America, which means that approximately 30–35 percent of homeowners cannot subscribe to a cable modem service. In addition, cable service usually bypasses businesses in shopping malls, strip shopping centers, and stand-alone business structures. This means that small businesses may have difficulty subscribing to a cable modem service.

Because power lines are routed into every home and business, BPL represents a ubiquitous offering that has the potential to reach every home and business. This means that for approximately 30–35 percent of homes that are currently bypassed by cable TV, BPL's competitors are primarily dial-up via the PSTN or DSL. As we will note when we discuss DSL technology, there are certain distance limits which typically removes it from being available for use by a significant percentage of homes and offices.

Even when cable modem service is available to the home or office, one needs to consider the layout of cable outlets in the facility. For example, most homes have cable TV outlets in the den, kitchen, and perhaps one or two bedrooms. If the homeowner has a built-in desk in the kitchen, which would be a good location for a home computer, chances are relatively high that the cable TV outlet is located on a wall by an island or another location that, although suitable for supporting a television, could not be easily used for a cable modem connection. Thus, the use of a cable modem commonly requires the installation of a new cable outlet in the home or office. This also means that someone must be in the home or office to allow the cable technician to gain access to the structure to install the new outlet. Thus, there is typically a degree of inconvenience associated with the initial installation of cable modem service that may not be present if one was to select a BPL service to obtain a high-speed Internet access capability.

Another difference between BPL and cable modem service that warrants discussion is the data rate supported by each technology. As mentioned at the beginning of this chapter, BPL technology commonly supports data rates between 500 kbps and 3 Mbps. In comparison, many cable operators now offer a fast access data rate up to approximately 5 Mbps. Although this data rate is significantly above the highest data rate obtained from existing BPL field trials, it should be mentioned that one pioneer of BPL — Media Fusion — amazed both potential customers and Internet Service Providers with a promise of achieving data rates up to 2.5 Gbps. Although the company has not yet been able to achieve anywhere near its projected data rate, it is currently working with the technology to increase the data rate. Thus, the maximum BPL data rate can be considered a work in progress.

Current cable modem rates vary between $24 per month for approximately 500-kbps service to $49 per month for high-speed access. This rate is approximately the cost associated with several BPL trials that will be discussed later in this book.

DSL

Digital Subscriber Line (DSL) technology currently represents the second most popular method for obtaining a high-speed Internet access capability in North America, serving about half the number of cable modem users. Although DSL subscribers exceed ten million, some key limitations preclude the service offering from becoming ubiquitous. One key limitation is the fact that the subscriber must be located no further than approximately 18,000 ft, or about 3 miles, from a telephone office. This limits the number of homes and offices capable of being supported by DSL service. In comparison, the fact that every home and office has electrical service means that BPL has the potential to be ubiquitous whereas DSL does not.

For homes and offices that can be serviced via DSL, this service is similar to cable service with respect to the fact that telephone outlets may not be in close proximity to the area where a computer will reside. However, because a typical home has slightly more telephone outlets, the ability to connect computer equipment to the telephone company DSL service is probably easier than establishing a cable modem connection. In comparison, DSL is a bit more difficult than when BPL is used because a home has more electrical outlets than telephone outlets.

Until recently, most DSL installations required a telephone company technician to visit the subscriber location to install filters required to separate data from the 300-Hz to 3,300-Hz voice conversation frequency.

Recently, a new generation of DSL modems includes adaptive filters that alleviate the necessity for a telephone technician to visit the subscriber's premises. In fact, several telephone companies now mail the DSL modem to new subscribers because it can be easily installed. Thus, DSL is probably equal in customer convenience and ease of setup to BPL modems provided to customers during field trials.

Concerning data transmission capability, DSL's maximum achievable data rate decreases as the distance between the subscriber's premises and the serving telephone office increases. Currently the maximum data rate obtainable through the use of the most popular type of DSL service, ADSL, is approximately 1.5 Mbps. However, the average DSL user more than likely experiences a maximum data transmission rate of approximately 1 Mbps. It should also be noted that some telephone companies offer a low-cost data transmission DSL service that is typically limited to 256 kbps. Although approximately five times faster than dial-up PSTN usage, the so-called "DSL Lite" represents the lowest high-speed Internet access technology.

The monthly cost of DSL services when first offered a few years ago was approximately $40 per month. Since its initial offering, large communications carriers have reduced the monthly cost of the service to the prices charged by cable modem providers to be more competitive.

Satellite

Currently, Internet access via satellite represents a niche market, with less than a few percent of all Internet access occurring using this technology. Satellite Internet access is commonly used in rural areas where neither cable modem nor DSL services are available. Although Dish Networks spent approximately a billion dollars for the creation of a satellite that was to provide a significant increase in Internet access, according to the *Wall Street Journal*, the company decided to use the satellite for video operations, which put a damper on its expected growth in the data services area.

One of the more unusual aspects associated with satellite Internet access is its integration with a wireless mesh network to provide high-speed Internet access to small communities. The wireless mesh network consists of a series of IEEE 802.11a, b, or b/g network adapters typically housed in a box, which in turn is mounted on light poles that relay transmissions to and from one or more access points. By connecting the access points to satellite stations linked to the Internet, the mesh network provides users within transmission distance of different mesh-enabled devices with access to the Internet.

Each station that participates in the mesh functions as a simple router, forwarding packets that are not destined for the station. Stations can enter and leave the mesh dynamically, with the routing protocol used by stations dynamically adjusting to such changes.

Because a wireless mesh network provides the ability for a group of users to access the Internet, it can be more cost effective than the use of individual connections. This economic advantage becomes possible when the cost of the high-speed Internet connection is amortized over a sufficient base of users. Although the BPL trials in effect when this book was written did not use a wireless mesh network to support a group of customers, there is no reason why the technology could not do so. Thus, the economic advantage associated with the connection of a wireless mesh network to a satellite station in comparison to the cost of individual high-speed Internet access would be eliminated if the mesh network was connected to an electric utility company power line that provided high-speed Internet access.

On an individual basis, a comparison between the use of a satellite and BPL for Internet access is anything but simple. Some satellite providers offer an integrated dish that enables a subscriber to obtain both video and data transmission capability, whereas other providers require separate satellite dishes. Because the use of satellite for high-speed Internet access is primarily located in rural areas and represents a few percent of all high-speed Internet access methods, it is unlikely to have any material effect on BPL once the latter is offered in rural areas.

WiMax

In concluding our discussion of competitive Internet access technologies, we will turn our attention to WiMax, an acronym that stands for worldwide interoperability for microwave devices. WiMax represents a wide area networking technology that can be used to transmit broadband signals over the air at distances up to approximately 30 miles. Although WiMax was standardized by the IEEE as the 802.16 standard in April 2002, its original specification resulted in several deployment problems. As we will note, a newer version of WiMax, referred to as the IEEE 802.16a standard, is presently being examined for use in several upcoming field trials.

WiMax has some similarities to wireless LANs (local area networks); however, it is also significantly different from the IEEE series of 802.11 wireless LAN standards. Concerning similarities, like the 802.11-compliant products, WiMax involves the use of client stations that use antennae to communicate with a centralized station. That centralized station is referred to as a central radio base station under IEEE 802.16 terminology and is designed to provide an alternative to cabled access networks, such as coaxial-based systems operated by your cable company in which you use a cable modem to access the Internet and a DSL for Internet access commonly offered by your local telephone company.

Although the original 802.16 standard was defined for use in the 10- to 66-GHz frequency band that represents spectrum available on a global basis, such high frequencies represent a significant deployment problem. This problem results from the fact that high frequencies have short periods, which restricts transmission to line-of-sight operations. Although 802.16 devices might be suitable for rural areas, where there are no tall buildings or other obstacles, field trials are typically performed to obtain a large base of users. Because this would require field trials to occur in metropolitan areas, the line-of-sight problem associated with the 802.16 standard in effect resulted in a delay in fielding such trials. Recognizing the line-of-sight problem associated with the 802.16 standard, the IEEE developed an extension to that standard, which became the 802.16a standard. This new standard defines an extension of WiMax to operate at lower frequencies, in the 2- to 11-GHz band, to include operations in both licensed and unlicensed frequency bands.

WiMax supports point-to-point and point-to-multipoint transmission methods similar to the IEEE series of 802.11 wireless LAN standards. The key difference between the two is the fact that wireless LANs can communicate only within a limited distance of a few hundred feet within a building to approximately a thousand feet outdoors, whereas WiMax supports transmission distances up to 30 miles.

At the time this book was prepared, several field trials based on the 802.16a standard were either being conducted or planned for future operation. One of the reasons for the delay in field trials was a lack of 802.16a chips. Although Intel had expressed keen interest in the technology, it may be late 2006 before 802.16a chip production occurs on a scale sufficient to result in cost-effective products.

Although WiMax can be viewed as a competitor to BPL, it can also be viewed as a supplemental technology. This results in the fact that the deployment of 802.16 central radio base stations in rural areas would require a mechanism to connect each base station to the Internet. Because power lines reach virtually every community regardless of their size or location, in the future it may be possible to use WiMax to serve clusters of homes and offices in rural areas and use BPL technology to connect central radio base stations via power lines to the Internet. This may be especially true for rural areas, where cable and DSL are not offered due to the low density of potential customers within a geographic area. In such situations the use of WiMax connected to a high-speed data transmission system provided through the use of low- and medium-voltage power lines could represent a mechanism to provide economical Internet access to homes and offices spread out over a large area.

Chapter 2

Power Line Operations

One of the most basic errors an author can make is to assume readers are familiar with the underlying technology associated with the topic of the book. After making this mistake a few times when I was still "wet behind the pen," I vowed not to make this mistake again. Thus, the purpose of this chapter is to acquaint readers with the operation of electrical power lines. To accomplish this goal, I will first focus our attention on understanding basic electricity; examining how circuits operate; the difference between alternating current (ac) and direct current (dc); the significance of the volt (V), ampere (I), ohm (Ω), and watt (W); as well as the manner by which electricity reaches the home or office. Because many emerging power line operation standards as well as compatibility issues require knowledge of specific terminology, such as the decibel milliwatt (dBm) and carrier frequency expressed in terms of kilohertz (kHz) or megahertz (MHz), we will also discuss the significance of those terms to ensure all readers have the same minimum level of background information required to benefit from the information provided in this book.

2.1 Understanding Electricity

In this section we will focus our attention on obtaining an appreciation of electricity. Commencing with a review of atoms and their electrons, we will discuss conductors, how current flows in a wire, the manner by which electricity is generated, basic circuit measurements, and the difference between alternating current and direct current.

Atoms and Electrons

Electricity represents a form of energy that occurs due to the flow of electrons. Every atom, which represents a microscopic structure found in all matter, contains one or more electrons that have a negative charge. Those electrons rotate around the atom similar to the manner by which planets rotate around the sun. In some materials, such as rubber, plastic, glass, and even air, electrons are tightly bound to the atoms. Because the electrons in effect do not move, they cannot conduct electricity and are referred to as electrical insulators. In comparison, most metals have electrons that can be easily detached from their atoms. Such electrons are referred to as free electrons and provide the ability for electricity to flow through different types of material. Copper, gold, silver, aluminum, and other metals that have electrons that can be easily detached from their atom are known as electrical conductors because they conduct electricity. Although electricity requires a conductor to move, it also requires something to make it flow through the conductor. That something is an electric generator.

Electric Generators

An electric generator converts mechanical energy into electrical energy based on the relationship between magnetism and electricity. Physics instructors often show their classes a small experiment during which a wire is moved across an electric field, resulting in the flow of an electrical current in the wire. A similar but reverse experiment sometimes used by physics instructors is to move electrons through a wire and observe how their flow results in the creation of a magnetic field around the wire. An example of the latter is illustrated in Figure 2.1, where a battery is used to generate the flow of electrons when the switch is closed. At that time the compass needle will move due to the magnetic field created by the flow of electrons. As we will note, the relationship between magnetism and the flow of electricity paved the way for the construction of power plant generators, which, in effect, to borrow an old tune, will "light up our lives."

Returning our attention to Figure 2.1, we can use that illustration as a mechanism to state the "right-hand law" discussed in many physics classes. That is, first clasp together the four fingers on your right hand

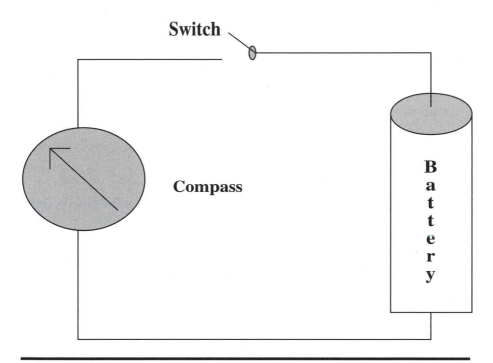

Figure 2.1 **The relationship between electricity and magnetism can be observed by closing the switch and viewing the change in the direction of the needle of the compass.**

and move them inward toward your palm. As you do so, extend your thumb upright. You can use your hand to view the relationship between the flow of current and the induced electromagnetic field. That is, the thumb indicates the direction of the flow of electricity, whereas the position of the four fingers indicates the direction of the electromagnetic force. For example, if the thumb is pointing upward, then the electromagnetic force flows counterclockwise. In comparison, if the thumb points down to indicate a reversal of polarity, that causes current to flow in the opposite direction; then the electromagnetic force flows in the clockwise direction. Unfortunately, if you are left handed, you should not favor that hand but instead use your right hand to comply with the right-hand rule. Now that we have an appreciation for the relationship between an electromagnetic force and the flow of current, let's turn our attention to how generators operate.

The large generators used by an electric utility have a stationary conductor formed in a ring shape. A magnet attached to the end of a rotating shaft is located inside the stationary conducting ring, and the conducting ring is wrapped with wire. As the magnet rotates, it induces a small electric current as it passes each section of wire. Cumulatively, the series of electric currents results in one current of considerable size. The shaft positioned within the stationary conducting ring is driven by a turbine whose blades usually are turned by water or steam. Thus, let's discuss the role of the turbine in the electric generator process.

Turbines

The most basic type of turbine is a water-powered device, with water from a dam falling over its edge, causing the blades of a turbine to spin. A second type of turbine is a fossil-fueled steam turbine, in which natural gas, oil, coal, or nuclear power is used to heat water in a boiler to produce steam. The resulting steam is pressurized and flows through the turbine, resulting in its blades spinning, which in turn spins the shaft that generates electricity. Because the turbine and generator are closely related, many manufacturers, such as General Electric, produce what is referred to as a turbine generator.

Circuit Measurements

The magnitude of the generated current resulting from the flow of electrons is referred to as amperage, which is measured in amps and denoted by the symbol I. In comparison, the pressure used to push electrons is referred to as voltage, which is measured in volts and denoted by the symbol V.

As electrons flow in a conductor, they encounter resistance. Resistance represents the ratio of voltage to current and is measured in ohms; 1 ohm is defined as the resistance when 1 volt is applied to a material and the current is 1 amp. Resistance is denoted by the symbol R. The relationship between voltage, current, and resistance is referred to as ohm's law, where:

$$R \text{ (ohms)} = \text{voltage/current} = \text{volts/amps} = V/I$$

Using ohm's law, we need to know only two of three measurements to compute the third. For example, assume we measure the current in a circuit to be 5 amps when 120 volts is applied. From ohm's law we obtain:

$$R = V/I = 120 \text{ V} / 5 \text{ I} = 24 \text{ ohms}$$

Similarly, if we know the resistance of a circuit is 24 ohms and we measure the current flowing through the circuit, we can compute the voltage. That is, again from ohm's law, we have:

$$R = V/I \text{ or } V = I * R$$

Then,

$$V = 5 \text{ amps} * 24 \text{ ohms} = 120 \text{ volts}$$

Circuits

Although electrical circuits can become quite complex, we can obtain an appreciation for how they operate by focusing our attention on their common components, using as a frame of reference the simple series circuit shown in Figure 2.2. Regardless of the type of electrical circuit, all circuits have several factors in common. First, the source of electricity, which is shown as a battery in Figure 2.2, has two terminals: a positive terminal and a negative terminal. Second, the source of electricity, which can be a simple battery or can result from a generator located hundreds of miles away, pushes electrons at a certain voltage. In the example shown in Figure 2.2, a typical AA battery will push electrons out at 1.5 volts. As electrons flow from the negative terminal to the positive terminal through the conductor, the path forms a circuit.

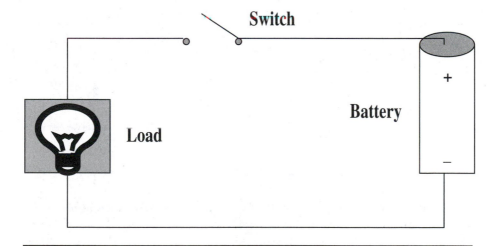

Figure 2.2   All circuits have a source of electricity, a load, and two wires used to transport electricity between the source and the load.

The load on the circuit can be a light bulb, television, stereo, or motor and will be powered by the electricity flowing through the circuit.

From the preceding example you might be a bit puzzled by the statement "as electrons flow from the negative terminal…" because many people view current flow as occurring from the positive terminal of a battery. For many years it was assumed that current flowed in this manner, until the development of equipment that showed the reverse was true. Thus, current flows from the negative terminal to the positive terminal and not vice versa.

Electrical Power

Power can be considered to represent the rate of doing work. In an electrical circuit, power is the rate of using energy and is measured in watts, denoted by the symbol W.

Power is also the product of volts and current in a circuit, such that:

$$\text{Power (watts)} = \text{voltage} \times \text{current} = \text{volts} \times \text{amps} = V * I$$

In North America, the power outlets in the walls of homes and offices are configured so that they are delivering 120 volts. If you locate a portable space heater and view its label, you might note that it operates at 120 volts and draws 10 amps. This means that its power in watts is

$$\text{Power (watts)} = 120 \text{ volts} \times 10 \text{ amps} = 1,200$$

If, instead of a space heater, you locate a 60-watt light bulb and plug it into a lamp, which in turn is connected to a 120-volt receptacle, you can use the preceding equation to compute the current that flows in the circuit that illuminates the light bulb. Because $P = V * I$, then:

$$I = P/V = 60/120 = .5 \text{ amp}$$

Thus, a 60-watt light bulb placed in a 120-volt socket draws .5 amps when illuminated.

Because electricity is consumed over time, power companies measure consumption in terms of power used over a period of time. For example, a 60-watt bulb lit for a period of 8 hours results in the use of 60 watts times 8 hours, or 480 watt-hours (Wh). Because the watt represents a relatively small amount of power, electric utilities bill customers based on the number of kilowatt-hours of electricity consumed; the kilowatt-hour (kWh) represents 1,000 Wh. Exceptions to the preceding are factories and other large consumers of electrical power. Such consumers may receive their monthly utility bills indicating consumption in terms of megawatt-hours (MWh); the megawatt-hour represents 1,000 kWh or 1,000,000 Wh.

Electric rates in the United States commonly vary from approximately 5 cents per kilowatt-hour to 15 cents per kilowatt-hour. Typically, the kilowatt-hour cost is lower in locations where electricity is generated through the use of water flowing across different types of barriers, referred to as hydro-electric generation. Of course, when an area that depends on hydro-electric generation experiences a drought, the utility is forced to import more expensive fossil-fuel or nuclear-generated electricity.

If we use an average of 10 cents per kilowatt-hour for electricity, we can note the bargain it provides in comparison to the functions it allows. For example, a 100-watt bulb left on during an 8-hour day consumes 100 watts times 8 hours, or 0.8 kWh of electricity. At a cost of 10 cents per kilowatt-hour, it costs just 8 cents to light up a desk throughout the workday. Because you cannot buy much in a candy store for under a dime and probably consume your purchase in less than a minute, the ability to light your desk for 8 cents during the workday is truly a bargain.

For a second example of the bargain of electricity, consider the 25-cubic-foot refrigerator installed in most homes. A modern 25-cubic-foot refrigerator will consume approximately 575 kWh of power during its full year of operation. At a cost of 10 cents per kilowatt-hour, this works out to $57.50 to operate your refrigerator throughout the year.

Direct Current versus Alternating Current

Batteries, solar panels, and fuel cells generate current that always flows in the same direction between their terminals. Thus, these types of power generators produce direct current (dc).

Because power plants generate electricity by moving a shaft that turns a magnet inside a stationary conducting ring, the direction of current periodically reverses, or alternates, resulting in power referred to as alternating current. In North America, current reverses direction 60 times per second, whereas in Europe, generators produce power that alternates the direction of current 50 times per second. This explains why we refer to current in North America being produced at

60 cycles per second (cps), whereas current in Europe is produced at 50 cps.

The major advantage of alternating current is the fact that it is very easy to change the voltage of the power produced by a generating plant. As we will note in the second section in this chapter, a transformer can be used to increase or decrease voltage. The key reason for the use of ac in homes and offices is one of economics. It is more economical for a power plant generating several million watts (MW) of power to transmit a high voltage with low amperage instead of a low voltage with high amperage. For example, a power plant capable of generating 1 million watts of power could transmit 1 MV at 1 amp. Sending 1 amp requires only a thin wire. In comparison, transmitting 1 M amps at 1 volt would require a very large-diameter wire and would result in a considerable loss of power due to the heat associated with moving a million amps through a large conductor. Thus, it's more economical to transmit power at high voltage and relatively low amperage.

Power plants use transformers to convert lower-voltage-generated alternating current to very high voltages for transmission. Because high voltage can literally fry both humans and equipment, a series of transformers is used to reduce the voltage that flows from main feeder lines onto branch lines, whereas other transformers reduce the voltage on branch lines down to the 120 volts that typically enters the home or office.

Ground

Ground, as the term implies, refers to a connection to the earth. A ground acts similar to a reservoir of charge and functions as a mechanism to prevent shock. On a transmission line you can see electric poles where a bare wire is connected from the upper portion of the pole to a coil placed in the base of the pole. The coil is in direct contact with the earth, which provides a ground. If you carefully look at a series of electric poles in your neighborhood, you can see the ground wire running between poles as well as attached to the coil at the base of each pole.

Communications Measurements

We need to review communications measurements for two basic reasons. First, we need an understanding concerning how we can compare

Table 2.1   Common Prefixes of Powers of 10

Prefix	Meaning
milli	1/1000 (thousandth)
kilo	1,000 (thousand)
mega	1,000,000 (million)
giga	1,000,000,000 (billion)

and contrast signals used for communications that flow over power lines. Second, understanding communications measurements provides us with the background necessary to understand the relationship between communications over power lines and potential interference caused by such communications. Thus, in concluding this initial section in this chapter, we will turn our attention to several measurements commonly associated with communications. We will first review the prefixes for the power of 10 because they represent a key to understanding many terms associated with electrical power generation and consumption.

Powers of 10

As a refresher for those readers who may be a bit rusty remembering prefixes of the powers of ten, let's examine the entries in Table 2.1. That table lists four common prefixes and their meaning. The first prefix, milli, represents one thousandth. Thus, the term milliwatt would represent a thousandth of a watt. The second term, kilo, represents a thousand. As we noted earlier in this chapter, kilowatt represents a thousand watts. The two remaining prefix terms, mega and giga, represent a million and a billion, respectively.

When we discuss power consumption in terms of kilowatt-hours, the use of that term is more appropriate to homes and small offices. In comparison, large organizations that consume a lot of power during the month may receive a power bill indicating consumption in terms of megawatt hours of electricity.

Power Measurements

The invention of the telephone resulted in the need to define the relationship between the received power of a signal and its original power. Initially, this relationship was given the name bel (B) in honor of Alexander Graham Bell, the inventor of the telephone.

The bel

The bel uses logarithms to the base 10 to express the ratio of power transmitted to power received. The resulting gain or loss for a circuit is given by the following formula:

$$B = \log_{10} P_O/P_I$$

where B is the power ratio in bels, P_O represents the output or received power, and P_I is the input or transmitted power.

The rationale for the use of logarithms can be traced to the manner by which humans hear. That is, our ears perceive sound or loudness on a logarithmic scale. For example, if we estimate that a signal doubled in loudness, the transmission power actually increased by approximately a factor of 10. A second reason for the use of logarithms in power measurements is the fact that changes to a signal in the form of a power boost from an amplifier or a signal loss due to resistance are additive. Thus, the ability to add and subtract when performing power measurements based on a log scale simplifies computations. For example, a 15-B signal that encounters a 10-B loss and is then passed through a 15-B amplifier results in a total signal strength of 15 − 10 + 15, or 20 B.

For those readers who are a bit rusty concerning the use of logarithms, note that the logarithm to the base 10 ($\log_{10}$) of a number is equivalent to how many times 10 is raised to a power to equal the number. For example, $\log_{10}$ 100 is 2, $\log_{10}$ 1,000 is 3, and so on. Because output or received power is normally less than the input or transmitted power, the denominator in the preceding equation is normally larger than the numerator. To simplify computations, note that a second important property of logarithms is the fact that the reciprocal of a logarithm is equal to its negative value. That is,

$$\log_{10} 1/X = -\log_{10} X$$

As an example of the use of the bel for computing the ratio of power received to power transmitted, assume that the received power is one-tenth the transmitted power.

Then,

$$B = \log_{10} 1/10/1 = \log_{10} 1/10$$

Because, from the properties of logarithms,

$$\log_{10} 1/X = -\log_{10} X$$

we obtain

$$B = -\log_{10} 10 = -1$$

From the preceding computation you will note that a negative value indicates a power loss, whereas a positive value would indicate a power gain.

Although the bel was used for many years to categorize the quality of transmission on a circuit, a more precise measurement was required by industry. This need resulted in the adoption of the decibel as the preferred power measurement.

The Decibel

The decibel (dB) represents the standard measurement used today to denote power gains and losses. The decibel is a more precise measurement because it represents one-tenth of a bel. The power measurement in decibels is computed as follows:

$$dB = 10 \log_{10} P_O/P_I$$

where dB is the power ratio in decibels, P_O is the output or received power, and P_I is the input or transmitted power. Returning to the preceding example covered during the discussion of the bel, where the received power was measured to be one-tenth of the transmitted power, the power ratio in decibels becomes

$$dB = 10 \log_{10} 1/10/1 = \log_{10} 1/10$$

Because $\log_{10} 1/X = -\log_{10}X$, we obtain

$$dB = -10 \log_{10} 10 = -10$$

Decibel above 1 mW

In concluding our examination of communications power measurement terms, we need to note that the bel and decibel do not indicate power. Instead, they represent a ratio or comparison between two power values, such as input and output power. Because it is often desirable to express power levels with respect to a fixed reference, it is common to use a 1-mW standard input for comparison purposes. In the wonderful world of communications testing, the 1-mW signal occurs at a frequency of 800 Hz. The use of a 1-mW test tone occurring at 800 Hz results from the evolution of early telephone systems. From measurements occurring at that time, 1 mW of power was equivalent to the power generated by an average person talking on the telephone and 800 Hz represented the average frequency of a human conversation. To ensure that you do not forget that the resulting power measurement occurred with respect to a 1-mW input signal, the term decibel-milliwatt (dBm) is used. Here, decibel-milliwatt represents

$$dBm = 10 \log_{10} \text{output power/1 mW input}$$

The m in dBm serves to remind us that the output measurement occurred with respect to a 1-mW test tone. Although the term decibel-milliwatt is used in most literature, in actuality it means "decibel above 1 mW." Thus, 10 dBm represents a signal 10 dB above, or bigger than, 1 mW, whereas 20 dBm represents a signal 20 dB above 1 mW, and so on. Because a 30-dBm signal is 30 dB, or 1,000 times larger than a 1-mW signal, this means that 30 dBm is the same as 1 W. Thus, we can use this relationship to construct a table indicating the equivalence between values expressed in terms of power in watts and power in decibel-milliwatts. Table 2.2 presents four key relationships.

We can better understand the relationships shown in Table 2.2 by commencing our observation with the previously noted fact that 1 W is equivalent to 30 dBm. Then we can note that 1 mW is a thousandth

Table 2.2 Relationship between Watts and Decibel-Milliwatts

Power in Watts	Power in Decibel-Milliwatts
.1 mW	–10 dBm
1 mW	0 dBm
1 W	30 dBm
1 kW	60 dBm

of a watt. Because 0 dBm means that output power equals input power, the only way to obtain a 0 dBm power level is for the output power to be 1 mW. Thus, the output power divided by a 1-mW input power value produces a 0-dBm power level and indicates that 1 mW is equal to 0 dBm. If you perform the previously mentioned computation, it's important to remember that we are working with logs. Thus, the equation for the decibel-milliwatt measurement is

$$dBm = 10 \log_{10} P_O/1 \text{ mW}$$

When we set P_O equal to 1 mW, we obtain

$$dBm = 10 \log_{10} 1$$

Because $\log_{10} 1$ is equivalent to determining that 10 raised to the power of 0 results in 1, the value of $\log_{10} 1$ is 0. Thus, dBm must be 0 for the output power to equal 1 mW. Now let's examine the .1 mW value. A .1 mW value is one-tenth of a milliwatt. Similarly, a −10 dBm value represents a tenth of 0 dBm. Thus, −10 dBm is equivalent to a power of .1 mW. We can now examine the last entry in Table 2.2, which indicates 1 kW is equal to 60 dBm. Because 1 kW is 1,000 times a watt and 60 dBm is 1,000 times greater than 30 dBm, then 1 kW is equal to 60 dBm.

2.2 Electricity Distribution

Most homeowners and office workers take electricity for granted; unless a storm knocks out our power lines, we really do not place much emphasis on how power is distributed to our location. In this section we will not wait for a sudden storm or for an automobile or truck to crash into a power company facility to impede the flow of electricity to appreciate how it arrives at our door. Instead, we will turn our attention to the manner by which electricity travels into our home or office.

The Power Plant

Electricity begins its travel to our home or office at a power plant. That plant uses water, coal, natural gas, or nuclear fuel to generate steam, which is used to make a turbine spin. The spinning of the turbine in turn generates electricity. In the case where water is used, typically a dam is constructed to gather water into a narrow channel. As water flows into the channel it moves toward locks that can be raised or lowered and control the flow of water across the dam. As water flows through the locks, its pressure is used to spin the blades of a turbine. In effect, the force of gravity and the motion of the water cause the turbine to spin. When a nonrenewable fuel such as coal or natural gas is used, combustion is employed to heat water into steam. Similarly, when nuclear fuel is used, uranium control rods are inserted into a reactor to generate heat via a controlled nuclear fusion process, which is used to turn water into steam. For both nuclear- and non-nuclear-generated steam, the result is similar, with the steam pressurized and used to turn the blades of turbines connected to a shaft placed in a magnetic coil that generates electricity.

Transformers

Electricity is generated by a turbine installed at a power plant that is normally located at a distance from the majority of the plant's customers. For example, a large power plant located at 4 Corners, where the borders of Arizona, Utah, Nevada, and New Mexico meet, is located hundreds of miles from major population centers. To facilitate the flow of electricity over long distances, the power plant uses transformers to increase voltage so that electricity can flow on high-voltage lines with a minimal loss.

A transformer makes use of Faraday's law and the ferromagnetic properties of an iron core to either raise or lower ac voltage. As a refresher for readers who may have skipped a physics class that discussed magnetic properties, Faraday's law states that any change in the magnetic environment of a coil of wire will cause a voltage to be induced in the coil. By adjusting the area of an iron core and the number of turns of a wire on each side of the core, we can either raise or lower a secondary or output voltage with respect to the primary or input voltage.

Figure 2.3 illustrates the operation of a transformer. In this example the primary voltage is produced by the power plant generator. The output or secondary voltage is then transported via a high-voltage power line to a metropolitan area.

From Faraday's law:

$$V_s/V_p = N_s/N_p$$

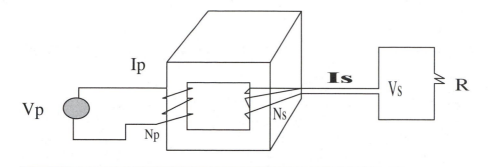

Figure 2.3 A transformer can be used to increase or decrease voltage.

where
V_s = secondary voltage
V_p = primary voltage
N_s = secondary number of turns
N_p = primary number of turns

Although a transformer can be used to raise or lower voltage, it cannot increase or decrease power. Thus, for an ideal transformer we have

$$P_p = V_p I_p = V_s I_s = P_s$$

where
P_p = primary power
P_s = secondary power
V_p = primary voltage
V_s = secondary voltage
I_p = current primary
I_s = current secondary

When the voltage is raised by a transformer, the transformer is referred to as a step-up device. Because the power cannot increase, when the voltage is raised, the current is proportionally lowered. Thus,

$$I_s = V_p I_p / V_s$$

In actuality, both the primary and secondary transformer circuits have resistance that affects the output of the device. Thus, the previously presented computations can be considered to represent an ideal transformer that has no resistance. Although transformers do indeed have resistance, for the purpose of this section, which explains the flow of electricity from a power station to a home or office, we can assume a perfect world and an ideal transformer. Thus, for the purpose of understanding the role of transformers, we need to note that the ratio of primary and secondary windings and primary voltage affects the resulting secondary voltage. If the number of primary windings (N_p) is greater than the number of secondary windings (N_s), then the secondary voltage (V_s) will be less than the primary voltage (N_p). If the number of secondary windings (N_s) is greater than the number of primary windings (N_p), then the opposite will be true. That is, the secondary voltage (V_s) will be greater than the primary voltage (V_p).

From the power plant's transformer, electricity is coupled onto high-voltage lines to enable power to flow relatively long distances. Where groups of customers are located, the high-voltage line is coupled to a transformer in a facility referred to as a substation. The transformer in the substation is a step-down device with a lesser number of secondary coils than primary coils. Thus, the substation reduces the voltage via a step-down transformer for distribution on power lines that can run either overhead or underground.

Types of Transformers

As the power lines reach a particular area where homes or offices are within close proximity, the voltage is further reduced by the use of smaller transformers. These transformers are easily recognized as circular containers mounted on utility poles or as small rectangular metal enclosures set on concrete pads. Figure 2.4 illustrates a pole-mounted transformer operated by Georgia Power in Macon, Georgia. Figure 2.5 shows a rectangular metal enclosure set on a concrete pad in my neighborhood. The metal enclosure contains a transformer whose output is placed onto power lines that run underground to homes in the neighborhood. Either type of transformer typically reduces the voltage provided by the regional transformer to 120 volts for use in the home or office.

Transformer Wiring

Because electricity flows in a circuit, there must be at least two wires routed from the last transformer that produced 120-volt output into the home or office. One wire represents a hot wire and the second represents a neutral wire. Older homes and offices that have only two

Figure 2.4 A utility pole–mounted transformer.

wires entering the facility can provide only 120-volt current. In comparison, more modern homes and offices commonly have three wires entering the facility. Two of those wires are hot, each carrying 120

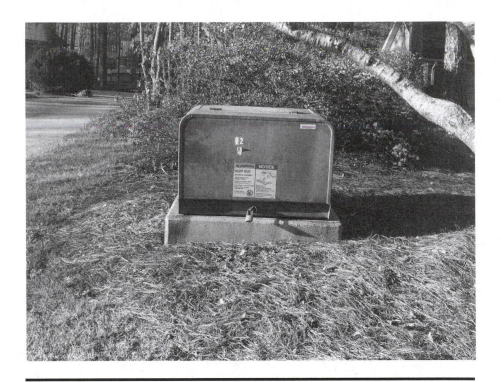

Figure 2.5 A metal enclosure used to house a transformer whose output flows onto underground power lines.

volts, and the third wire is neutral. Thus, the inbound voltage permits an electrician to provide circuits in the home or office that operate at either 120 or 240 volts.

Service Methods

Electric service to homes and offices is provided via two methods: overhead service and underground service. If the home or office has overhead service, the wires emerge from a transformer and are routed as an aerial line to a weatherhead mounted on the roof or side of the facility. The weatherhead is a conduit that allows wires from the transformer to be spliced into wires that are routed to an electrical meter. The meter measures how much electricity the house or office uses and provides information that affects the monthly bill. From the electric meter, wires are routed into a service panel, which distributes electricity throughout the home or office.

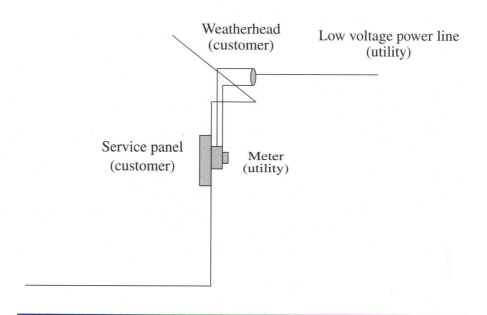

Figure 2.6   Relationship of power line, weatherhead, meter, and service panel when power is provided by an overhead line.

The Service Panel

The service panel represents the location in a home or office where electricity provided by a utility enters the facility for distribution within the facility. People often refer to service panels as fuse boxes or circuit breakers, although those terms are not technically correct.

Figure 2.6 illustrates the relationship between the utility line, weatherhead, utility meter, and service panel, indicating the ownership with respect to the customer and the utility. In examining Figure 2.6, note that although the service panel is shown inside the building, it can also be mounted outside; but doing so requires a weatherproof box, which adds to its cost as well as makes checking the panel a bit more difficult. Also note that an overhead utility line is shown providing power to the home or the office building. If an underground power line was used, the power line would be routed underground from a slab-mounted transformer to the edge of the home or office. At that location, the power line would be routed through a conduit in the form of a pipe into the electric meter. Thus, the weatherhead would not be needed.

Types of Service Panels

There are two types of service panels used in homes and offices. The first type, which is primarily associated with older homes and offices, is a fused service panel. This type of service panel, which was used in my first residence during the late 1960s, consists of a rectangular box into which is wired an electric line from the outside meter. Each of the circuits in the home is protected through the use of a circular fuse, which blows when too much current flows through it. Once blown, a fuse cannot be fixed; it must be replaced. Thus, homes and offices with fuse panels typically need a supply of fuses readily available as replacements when a fuse blows.

The second and more modern type of service panel is a breaker panel. This type of panel contains a series of circuit breakers that trip when they sense heat caused by an excess amount of current flowing through the breaker to the circuit protected by the device. When the circuit breaker trips, it shuts off power to the circuit and the power remains off until the breaker is physically reset. Most modern breakers will display a red tab when they are blown, as well as place themselves in the OFF position. Other breakers will simply move to the OFF position when they are tripped, whereas a push-button breaker will pop out when a trip occurs.

Regardless of the method used to indicate a trip condition, before resetting a breaker, you should first unplug any device on the circuit controlled by the breaker. Then, you can either flip the breaker to the ON position or, on a push-button device, push the breaker all the way in and then release it to perform a reset operation.

Operation

Figure 2.7 illustrates the breaker service panel in my current home. At the top of the box, power enters through two hot wires. Although you cannot see them, one of these wires is colored black and the other red; a white wire is used for the return. At the top of the service panel is the main breaker, which when flipped or repositioned turns off power to the entire home.

Depending on the location of a home or office, some electrical codes require the addition of a main disconnect location outside of the structure and other codes allow the main disconnect to be placed within the home or office. For either location, shutting off the power turns off the flow of electricity to all circuits routed from the panel through the home or office.

Figure 2.7   A breaker service panel.

Inside the Service Panel

Returning our attention to the service panel located in my current home, let's discuss the breakers shown in Figure 2.7. Note the two rows of breakers running vertically inside the panel. If you look carefully at Figure 2.7, you will note that on some rows there are two miniature breakers. Those minibreakers control low-amperage circuits whereas the normal-sized breakers control higher amperage circuits.

If you remove the metal panel in the circuit breaker box, which by the way should be done only by an experienced and licensed electrician, you will be able to see two hot wires, each carrying 120 volts, referred to as hot bus bars, routed to the main disconnect. A 120-volt breaker is attached to one wire, and a 240-volt breaker is attached to both wires. The hot wire for each circuit in my home is attached to each breaker, whereas a neutral bus bar has green or bare copper ground wires attached to the bar. In addition, you would note a thick ground wire attached to the neutral bus bar that represents the main ground wire. That wire is routed out of the service panel to either a cold-water pipe or rod, with the latter commonly required to be driven at least eight feet into the ground. If a cold-water pipe is used, because it leads into the ground via other pipes, it can also serve as a ground. In addition to the main ground, each individual circuit must be grounded. This is commonly accomplished by the use of a green or bare copper wire attached to the neutral bar.

Circuits

From the service panel each breaker or fuse protects a circuit routed throughout the home or office. Those circuits are referred to as branch circuits and can be subdivided into three categories: general purpose, small appliance, and individual circuits.

General Purpose Circuits

General purpose circuits are used to provide electricity to several receptacles used for lighting and small electrical appliances, such as fans, computers, and similar devices. As we noted earlier in this chapter, a single 100-watt light bulb draws approximately 1 ampere of current. Due to the low amperage requirement of lighting, a 120-volt general purpose circuit that commonly uses a 15- or 20-amp circuit breaker is wired to several receptacles, allowing a group of table lamps, fans, and a computer to use one branch circuit. Figure 2.8 illustrates the routing of wires from a circuit breaker to provide electricity for two receptacles and a switch that controls the flow of current to a light.

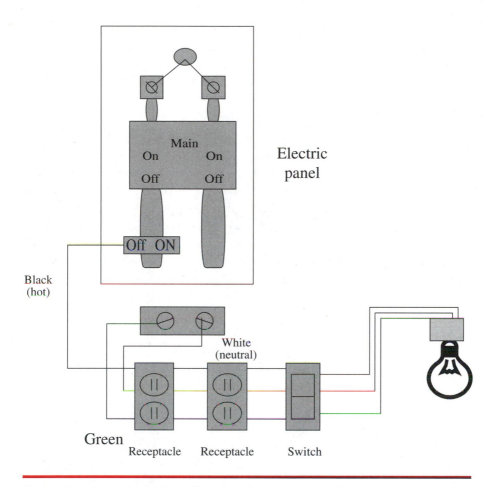

Figure 2.8   A typical 120-volt general purpose circuit.

During the late 1970s and early 1980s, initial products designed to use in-home electrical wiring as a mechanism to share what were then expensive peripheral devices at first were limited to working on a single electrical circuit. It wasn't until the 1990s that products emerged for home networking use that were capable of treating multiple circuits as a common network.

Small Appliance Circuit

A second type of circuit is used to power small appliances. Because such small appliances as microwaves, toasters, and mixers can draw a lot of current when turned on, each small appliance branch circuit is connected to only a few receptacles. Because appliances are primarily located in a kitchen or cafeteria area, the small appliance circuit is installed in those locations.

Individual Circuits

The purpose of an individual circuit is to provide electricity to a location where a lot of power will be consumed. Typically, individual circuits are used to provide power to a dishwasher, trash compactor, and similar appliances in the home. When large appliances, such as washers, dryers, electric ranges, and ovens are powered by individual circuits, such circuits provide 240-volt current.

Now that we have a general appreciation of electricity from its generation to its distribution within the home or office, we are ready to move on. Thus, grab a soda and your favorite munchies and let's continue our investigation into the use of power lines for communications.

Chapter 3

Data over Power Line Operations

In the first chapter we briefly noted that broadband over power line (BPL) operations can be viewed as operating similar to the manner in which Digital Subscriber Lines (DSLs) operate. In the second chapter we examined how electricity is generated and distributed. Using the information presented in those two chapters as a base or foundation of knowledge, we will now examine the different architectures associated with moving data over power lines as well as how data is actually transmitted over power lines. In doing so, we will first focus our attention on the different architectures used to provide a BPL capability over an electric utility infrastructure. As we review the architectures associated with different BPL systems, we will initially confine our discussion of transmission methods and modulation techniques to a brief description of the techniques and methods employed. However, in the second section in this chapter, we will focus our attention on several more popular transmission methods and modulation techniques used in several BPL field trials. As we describe and discuss each transmission method and modulation technique, we will also note the advantages and disadvantages associated with their use. Thus, the first section of this chapter will be focused on the system architectures associated with providing a BPL service and the second section will examine specific methods used to transport data over different architectures.

3.1 BPL System Architecture

Broadband over power line represents a technology that enables data to be transmitted over electric utility power lines. Subscribers residing in residential homes and offices commonly install a special type of modem that plugs into an ordinary electric wall outlet to obtain the ability to use electric utility power lines to access the Internet. Similar to DSL and cable modem subscribers, the BPL subscriber pays a monthly subscription fee and, in general, is not too concerned about the service provider's infrastructure until they either lose their communications capability or need information about the service to make an intelligent acquisition decision. For both of the preceding reasons, as well as the fact that this book was written to provide detailed BPL information, we need to examine both the network architecture used to provide the mechanism for a data transport facility as well as the transmission methods and modulation techniques used in several field trials in more detail.

Power Grid versus In-Building Wiring

BPL can be considered to represent a local area networking (LAN) technology due to the fact that the same transmission flows to each station residing on an electric circuit. Similar to a LAN, only the station with the destination address of a frame flowing on an electric circuit "reads" the frame flowing on the wire. In fact, CSMA/CA is used as the access protocol on most BPL systems. Thus, from a physical perspective, transmission over the power grid is similar to transmission in the home over in-building electric wiring because both types of wiring use a LAN access protocol. Although there is no clear distinction between the two, the Federal Communications Commission (FCC) has attempted to define one. When BPL operates within a building, it is referred to as in-building BPL. In comparison, when BPL occurs over the power grid, it is referred to as access BPL. Here the term "access BPL" can be considered in recognition of the fact that the power grid provides the access mechanism to the Internet. Viewed another way, in-building BPL can be considered to represent a LAN whereas access BPL represents a LAN technology viewed by the FCC as a wide area network (WAN). Although the underlying technologies are similar, the two are used and regulated differently. In this chapter we will focus our attention on access BPL, which we will, for simplicity, refer to as BPL. Later in this book we will discuss in-building BPL, which can be considered to represent a home networking technology and is standardized by the HomePlug Powerline Alliance.

Architecture Overview

Currently, BPL equipment vendors and electric utilities use several network architectures to provide data transmission through an electric utility power line infrastructure. Although considerable differences exist in the manner by which transmission occurs from the nearest point of the utility infrastructure into a subscriber's home or office, as we move upward in the infrastructure toward the power-generation plant, differences become negligible. For each architecture, BPL works by modulating high-frequency radio waves; however, as we will note, the transmission method and modulation technique can vary based on the architecture employed. Once modulated, radio waves are inserted into the electric utility grid at specific locations. Those modulated waves travel along the power lines until they reach a transformer or a location where the waves require amplification. Concerning transformers, they were designed to pass low frequencies near 60 Hz in North America and 50 Hz in Europe. Thus, they would appear as an open circuit to the passage of higher frequency signals and adversely affect the flow of data across power lines.

To provide telecommunications services, a utility must bypass the transformer, which can be expensive. Thus, one architecture we will examine in this section uses wireless LAN access points positioned at strategic locations to alleviate the need to bypass transformers that are installed by an electric utility to lower voltage on power lines routed into homes and offices. Another factor that governs network architecture relates to the manner by which neighborhood transformers service customers. As we will note in this section, in North America a neighborhood transformer may provide a low-voltage connection to only a handful of homes, whereas a similar transformer in Europe can provide service to a hundred homes. This difference can also affect the network architecture used to support communications over power lines.

Prior to discussing each specific BPL architecture, let's focus our attention on the general infrastructure of an electric utility. A utility transmits electric power at approximately 60 Hz in North America and 50 Hz in Europe, with varying voltages. The utility infrastructure can be divided into three voltage categories that correspond to the voltages placed on their power lines. Those voltage categories are high voltage (HV), medium voltage (MV), and low voltage (LV), resulting in the infrastructure of an electric utility consisting of high-, medium-, and low-voltage lines.

High-Voltage Lines

High-voltage lines are routed from a power-generating station to a substation. The high-voltage lines form the electric utility backbone power distribution system and commonly carry from 155,000 to 765,000 volts. Because that amount of power is very noisy, high-voltage lines are usually not used for data transmission. Instead, the excess capacity of traditional fiber-optic lines installed by electric utilities for monitoring and control purposes are used to bypass high-voltage power lines; however, some third-party vendors are developing equipment that could allow transmission over noisy high-voltage lines in the future.

Medium-Voltage Lines

Medium-voltage lines are routed from a substation to a neighborhood transformer. Because medium-voltage lines carry a more manageable 7,000 to 15,000 volts, they commonly form the backbone of an electric utilities data over power line infrastructure. In fact, some field trials involve connecting each utility medium-voltage line to the Internet instead of connecting the medium-voltage lines to fiber-optic lines that run in parallel with many high-voltage lines.

Low-Voltage Lines

From the neighborhood transformer, voltage is stepped down to the 120-volt, low voltage that is routed to and within homes and small businesses. Thus, it is the low-voltage line that is routed to the customer.

The Electric Utility Infrastructure

Figure 3.1 illustrates the general electric utility infrastructure that evolved during the past century. In examining the power line structure shown in Figure 3.1, note that the high-voltage lines commonly run directly from a power-generating facility to substations. The substations distribute power to neighborhoods via the use of medium-voltage lines, whereas the use of neighborhood transformers results in low-voltage

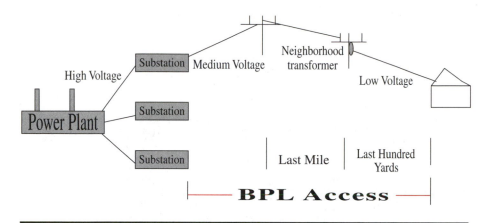

Figure 3.1   The general electric utility infrastructure.

lines spanning the last hundred yards or so into customer facilities. Also note that although the term "BPL access" is used to denote the route from medium- and low-voltage lines into subscriber homes and offices, it is also possible for the route to be extended to the route taken by high-voltage lines. However, as previously discussed, currently fiber-optic lines routed in parallel with high-voltage lines are used to transport data due to the noise on high-voltage lines, which makes them unsuitable for use as a data transport facility.

The electric utility infrastructure represents a configuration or power grid structure developed over a relatively long period of time. That time period for most well-established cities exceeds a century, with the power grid structure setup such that high-voltage lines are converted to medium-voltage lines and medium-voltage lines are then converted into low-voltage lines. Each conversion occurs through the use of transformers, which, as previously mentioned, commonly block communications signals transmitted at frequencies beyond that alternating cycle rate of the current carried on the power lines.

To alleviate the blocking effect of transformers on the passage of data requires the installation of a coupler between the medium- and low-voltage power lines. That coupler, which is referred to as an inductive coupler, will be discussed later in this chapter.

At utility poles within a neighborhood, another device, commonly referred to as a bridge, is installed. Depending on the vendor of this product, the bridge may include one or more communications functions. Table 3.1 lists the possible functions that can be performed by a bridge mounted on a utility pole containing a transformer that provides low-voltage power to a cluster of homes or offices.

Table 3.1   Bridge Functions

Support data routing
Manage subscriber information
Support DHCP[a] assignment of IP[b] addresses
Support encryption
Support symmetric data transmission to all electrical outlets in subscriber home or office
Support WiFi access

[a] Dynamic Host Configuration Protocol
[b] Internet Protocol

In examining the functions listed in Table 3.1, it's apparent that the bridge represents the heart of a BPL transmission system. This device, when supporting the IEEE 802.11 WiFi standard, enables a low-cost commodity such as wireless LAN adapters to be used by customers. When the bridge manages symmetric data transmission, it enables power line modems that plug into electric outlets in the home or office to directly communicate via the electrical wiring infrastructure. Thus, the bridge can be considered as the mechanism that controls the basic network architecture on which data flows to and from homes and offices.

Grid Problems

In addition to the need to bypass transformers, there are several additional problems associated with attempting to communicate over the electric infrastructure or grid of a utility. Those problems, which we will examine in this section, include the noise resulting from electric lines transporting high voltage, the built-in safety and fault-tolerance features embedded into the electric grid, and the manner by which homes and offices are connected to the grid.

Line Noise and Attenuation

High-voltage lines create noise and attenuate communications, resulting in most electric utilities either installing or using existing fiber-optic lines that are run in parallel along the high-voltage line paths that form a communications backbone for most utility power grids. The installation of fiber-optic lines usually occurred as a mechanism to provide the utility with a communications monitoring and control facility. Because advances in wavelength division multiplexing and dense wavelength division multiplexing increased the data communications capacity of optical fiber by several orders of magnitude, most electric utilities have a communications backbone in place that can be easily upgraded to support the communications requirements of their customers without having to install additional optical fiber.

Fault-Tolerance Equipment

A second problem associated with electric power grids involves the safety and fault-tolerance features built into the modern electric power grid. Those features can result in a high level of signal attenuation, which may require the use of additional amplifiers. Because amplifiers are costly, overcoming signal attenuation can be expensive.

Grid Connection Method

A third problem associated with the structure of the electric utility grid concerns the manner by which homes and offices are connected to neighborhood transformers. In Europe, most utilities have structured their power grids such that there are hundreds of homes and offices served by a neighborhood transformer. This structure enables an economic connection to be made from a communications backhaul, which is usually a fiber-optic line, to the low-voltage line below the neighborhood transformer. Thus, the European grid structure avoids the need to bypass the neighborhood transformer. In comparison, in North America the electric utility grid has only a dozen or fewer homes and small businesses per transformer. Thus, the backhaul must hook into the medium-voltage line, which then requires power line communications to bypass the neighborhood transformer.

In field trials conducted in the United States, the primary method used to bypass neighborhood transformers is commonly accomplished through the use of inductive coupling. In technical terms, inductive coupling represents the transfer of energy from one circuit to another due to the mutual inductance between the circuits. Inductive coupling is caused by the movement of current in a wire or on the metal of equipment. Similar to transformer theory described in the previous chapter of this book, moving currents generate magnetic fields, which create other moving currents in adjacent wires or conductors. Thus, the placement of a wire near the medium-voltage line behind the neighborhood transformer enables inductive coupling to pass the communications signal while bypassing the transformer. Although this method of bypassing the neighborhood transformer is relatively easy to accomplish, it is labor intensive and adds to the cost associated with providing a communications capability via the use of the electric grid.

Now that we have an appreciation for the general structure of the electric grid and the manner by which electricity flows to our homes and offices, let's turn our attention to the different network architectures used by BPL utilities in field trials. In doing so, let's first review the terminology associated with BPL equipment. As we will note, the development of BPL communications resulted in a new series of terms being associated with communications equipment developed to operate on electric power lines.

BPL Terminology

Previously we noted that the electric grid consists of high-, medium-, and low-voltage lines, with step-down transformers used to reduce voltages between each type of power line. In this section we will focus our attention on the BPL access facility, which typically extends from medium-voltage power lines through the neighborhood transformer that provides low voltage to home and small office electric customers. In the United States, medium-voltage lines typically transport between 1,000 and 40,000 volts, bringing power from an electrical substation to a neighborhood. In comparison, the low-voltage lines routed from the neighborhood transformer, which is also referred to as a distribution transformer, carry 120/240 volts for residential and small business use.

Backbone or Backhaul

In Figure 3.1 we examined the general electric utility infrastructure. In discussing the use of high-, medium-, and low-voltage lines, we noted that the fiber-optic lines many utilities previously installed for internal use can be easily upgraded to support customer transmission requirements. That fiber can then serve as the telecommunications backbone or backhaul data network for a utility offering BPL.

An electric utility has a well-defined power grid or distribution system. Thus, the addition of a communications capability results in the use of as much of the existing grid as technically and economically possible. Although the utility fiber-optic backhaul may be upgraded

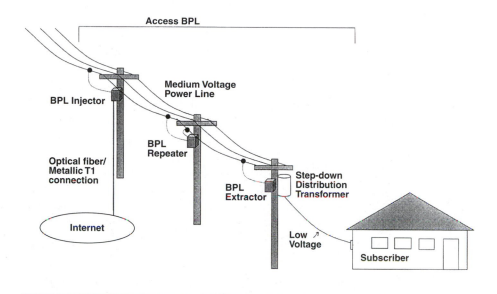

Figure 3.2   General BPL access infrastructure.

and used, in some cases it may be more economical to connect leased high-speed lines from third-party communications vendors to medium-voltage lines and bypass the backhaul. Thus, there is no single architecture suitable for all electric utilities. Instead, several architectures can be considered by most utilities.

Network Components

Similar to a conventional communications network, the transmission of data over power lines requires the use of several types of network components. Those components have specific terminology when referring to communications occurring over power lines. In this section we will examine the role of the injector, the repeater, and the extractor.

Figure 3.2 illustrates the basic structure of the BPL access infrastructure, which enables both electricity and communications to flow into the home and small business. As previously noted in this chapter, the medium- and low-voltage lines are referred to as "access BPL" by the FCC as a mechanism to separate transmission over power lines from in-building communications. We will use this illustration as a reference to denote the terms injector, repeater, and extractor associated with moving data over power lines as well as the terms single phase and three phase.

BPL Injector

In examining Figure 3.2, note the BPL injector on the left. The BPL injector, which is also known as a concentrator, is typically mounted in plain view on power poles located within a curb area in a community. A fiber-optic or metallic T1 line, which is connected to the Internet, is routed as an input connection to the injector, the latter commonly housed within a large gray or black box mounted on a utility pole. Inside the injector is a transmitter and receiver section as well as a signal converter. The transmitter and receiver operate on different frequencies, which in effect enables full duplex transmission over the power line. The injector converts the signal on the fiber or metallic T1 line into the signal format used for transmission over the medium-voltage power line. Typically, this action results in the generation of an orthogonal frequency division multiplexing (OFDM) signal, which consists of a series of carriers modulated using binary phase shift keying (BPSK) or another modulation method. The carriers are placed onto the power line in two blocks of frequency, with one block used for downlink transmission and the other to provide an uplink transmission capability. The BPL injector also couples the OFDM signal onto one phase of the medium-voltage power line. This is shown in Figure 3.2 by the connection of the BPL injector housing to one of the three lines on the leftmost telephone pole. Because the injector is bidirectional, it also converts OFDM signals flowing on the medium-voltage line into the format used by the Internet backbone.

BPL Repeater

The BPL repeater is installed approximately every 1,000 to 2,500 ft along the medium-voltage power line. The purpose of the repeater is to amplify and stabilize the signals transporting data over the medium-voltage power line.

Although most BPL literature refers to the device that amplifies and stabilizes signals as a repeater, this is not technically correct. In the wonderful world of communications, a repeater is normally thought of as a data regenerator, receiving a digital signal, discarding the signal, and then regenerating the signal so that any distortion is eliminated. Because OFDM and other modulation methods result in the transmission of analog signals, amplifiers are used. Although it is not technically correct, I will refer to devices that boost signal power as repeaters to remain consistent with the terminology used by BPL equipment vendors.

BPL Extractor

The third major device used under access BPL is the extractor. Extractors provide the interface between the medium-voltage power lines carrying BPL signals and homes and small businesses within the service area. BPL extractors are typically located at each low-voltage transformer, which provides low-voltage feeder lines to a group of homes. Some extractors include a built-in repeater, which boosts the signal strength to a sufficient level to enable transmission to occur through the low-voltage transformer.

In comparison, extractors without a built-in repeater relay the BPL signal around transformers via the use of couplers on the medium- and low-voltage lines. A third type of extractor interfaces with a non-BPL device, such as an IEEE 802.11 wireless LAN access point, which enables the BPL network to be extended to a group of customers.

Phase Lines

If you look carefully at Figure 3.2, you will note there are three wires used to deliver medium voltage on the electric grid. The use of three wires is based on three-phase electrical generation, which is now very common and is a more efficient use of commercial generators. As a refresher, electrical energy is generated by rotating a coil inside a magnetic field in large generators that have a very high capital cost. However, it is relatively simple and cost effective to include three separate coils on a single shaft instead of a single coil. Each coil is located on the generator's shaft but is physically separate and at an angle of 120° to each other. This placement results in the generation of three current waveforms that are 120° out of phase with each other but of equal magnitude, with each current flowing on a separate line; this is referred to as a three-phase distribution system. At the step-down distribution transformer, a single phase and neutral is normally routed to the subscriber.

Now that we have an appreciation for BPL network components to include the terminology of equipment used to provide communications over power lines, let's turn our attention to specific BPL network architectures commonly used in field trials.

BPL Network Architectures

OFDM-Based System

The first network architecture we will examine is based on the use of orthogonal frequency division multiplexing (OFDM) to distribute the

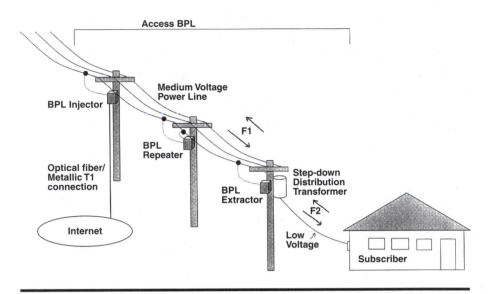

Figure 3.3   OFDM-based network architecture.

BPL signal over a wideband using numerous narrowband subcarriers. Because it uses OFDM, we will refer to this architecture as an OFDM-based system.

Figure 3.3 illustrates the OFDM-based system architecture. In examining the flow of data shown in Figure 3.3, we see the backhaul or backbone network connection to the Internet is converted into the OFDM signal format at the BPL injector. The output of the BPL injector is coupled onto one phase of the medium-voltage power line. In the opposite direction, BPL signals on the medium-voltage power line are converted to the format used on the backbone connection to the Internet.

As data flows on the medium-voltage lines toward the customer, extractors are used to route and convert data signals between the OFDM signal and the in-home BPL signal format. Depending on the distance between the BPL injector and the extractors it serves, one or more repeaters may be required to periodically boost the signal.

Both the injector and the extractors it serves share a common frequency band on the medium power lines. This frequency band, which is denoted as F1 in Figure 3.3, differs from the frequency used by the customer's in-house BPL devices, which is shown as F2 flowing on the low-voltage power lines. In actuality, if OFDM is used, the frequencies shown in Figure 3.3 represent blocks of spectrum. As repeaters (located periodically on medium-voltage lines but generally less than 2,500 ft from one another) are reached, each frequency block is converted into a new frequency spectrum. The new frequency block is used for approximately another 2,500 ft. Blocks of spectrum cannot be reused for several 2,500-ft legs on the power line. However, after two legs, a block can usually be reused without creating interference with a previously used frequency block.

To minimize contention, the Carrier Sense Multiple Access with Collision Avoidance (CSMA/CA) protocol is employed. Although BPL signals are coupled onto one phase line, if the signal is increased so that it can tolerate co-channel interference, it may be both practical and possible to implement two or three of these systems to operate independently of each other on adjacent medium-power lines.

OFDM–WiFi-Based System

A second network architecture that operates over power lines uses OFDM as its on-the-wire transmission method; however, instead of communicating over low-voltage lines into customer homes and offices, it uses an IEEE 802.11 wireless connection to the subscriber. Figure 3.4 illustrates the architecture of an OFDM–WiFi-based system.

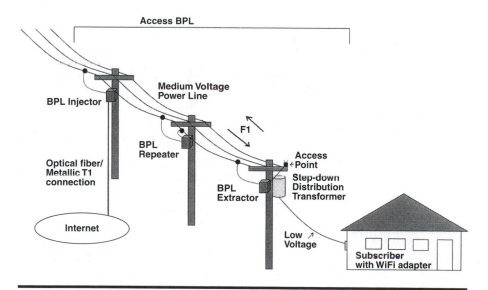

Figure 3.4   An OFDM–WiFi-based system.

In examining Figure 3.4, you will note that a BPL extractor is used to convert the BPL signal flowing on the medium-voltage line into an IEEE 802.11 signal. The latter provides a wireless signal to customers. In a number of field trials conducted during late 2004 through 2005, IEEE 802.11b access points were used to provide the wireless signal to a group of closely co-located customers. Although IEEE 802.11g access points will probably be used in the future due to the higher data rate and extended transmission distance provided by this newer standard, regardless of the WiFi method used, the important point of this architecture is the fact that it does not use BPL on low-voltage power lines. In addition, because an access point has a circular radius of transmission, it is capable of providing a broadband connection to a group of subscribers, which can reduce the service cost to the utility.

The OFDM–WiFi communications network architecture is overlaid on the same power line structure as previously illustrated in Figure 3.3. However, there are several differences between the two. First, the OFDM–WiFi system may use both repeaters and extractors to provide a connection to a WiFi access point. When a repeater provides a connection to a WiFi access point, it includes the capability of an extractor within its housing. A second difference between an OFDM system and a combined OFDM–WiFi system concerns the use of low-voltage lines. Because the OFDM–WiFi system uses access points, it does not use the low-voltage lines. This means that this network architecture does not have to employ a scheme to bypass the neighborhood transformer, which reduces the cost of the architecture. In addition, the WiFi access point can be used to support more than one customer, which further reduces the cost associated with this architecture. Although the OFDM–WiFi network architecture also couples BPL signals onto one phase of the medium-voltage power line, similar to an all-OFDM network architecture, it's possible to transmit over multiple phases on the three-phase medium-voltage power line. In fact, the next network architecture we will examine does this; however, it uses a different signaling method to transmit data over medium-voltage power lines. Because the next network architecture we will discuss uses direct sequence spread spectrum (DSSS), we will refer to this as a DSSS-based system.

DSSS-Based System

A third network architecture used to transmit data over power lines uses DSSS transmission. DSSS represents a transmission method

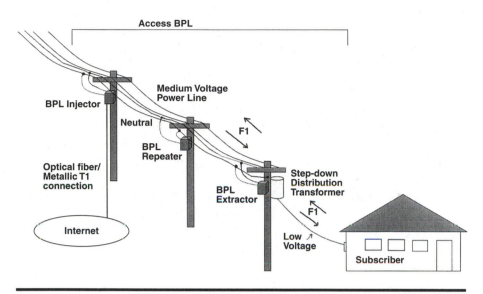

Figure 3.5 DSSS-based network architecture.

whereby a signal is modulated and spread over a range of frequencies, with multiple copies of each information bit transmitted. At the receiver, "majority rules" occurs. That is, if a "0" bit is transmitted on five frequencies and the receiver's demodulation results in four "0" bits and one "1" bit being demodulated, the receiver will assume the bit transmitted was a "0" bit.

Figure 3.5 illustrates in schematic form a DSSS network architecture under which that protocol is used to transmit BPL data over medium-voltage power lines. Under this network architecture, all users within a BPL cell share a common frequency band. Similar to the OFDM network architecture, CSMA/CA is used to minimize contention for a channel. Each BPL cell consists of an injector (concentrator) that provides an interface to the fiber or metallic connection to the Internet, a number of repeaters that compensate for signal loss as data flows over electric power lines, and neighborhood transformers that feed clusters of homes and offices.

Several field trials associated with the DSSS network architecture were implemented with slight modifications. Although some DSSS network architecture field trials coupled the BPL signal onto the medium-voltage power line using a pair of couplers on a phase and neutral line, one trial deployment resulted in the BPL service provider using two phases of the same run of a three-phase medium-voltage power line.

Summary

The three network architectures we previously reviewed are similar with respect to the use of the medium-voltage power lines. The OFDM network architecture uses different frequencies on medium- and low-voltage power lines for networking into a neighborhood and then extending the network to customers via low-voltage power lines. In the second network architecture, which represents a combination of OFDM and WiFi, BPL is limited to operating on medium-voltage power lines whereas WiFi provides the transport mechanism into the subscriber's premises. In the third network architecture, DSSS is used as a data modulation and transport method, with the same frequencies used on medium- and low-voltage power lines to provide a network capability directly into the user premises. Now that we have an appreciation for the three basic network architectures used to deliver data over the electric utility infrastructure, we will conclude this chapter by examining transmission methods, modulation techniques, and the network access protocol used to deliver information to power line customers.

3.2 Transmission Methods, Modulation Techniques, and Network Access

In this section we will cover a troika of interrelated topics. First, we will examine the operation of the two transmission methods used in conveying information over power lines: OFDM and DSSS. Then, as we discuss each transmission method, we will also discuss applicable modulation methods.

The third major topic we will discuss in this chapter is the manner by which multiple users located in different residences gain access to the power line network. As previously noted in this book, BPL uses the CSMA/CA protocol, which can be considered to represent a version of Ethernet that uses collision avoidance instead of Ethernet's collision detection scheme.

Transmission Methods

BPL field trials resulted in the use of two transmission methods to convey information over power lines. One method, referred to as orthogonal frequency division multiplexing (OFDM), can be traced to the mid-1980s when it was used in the first high-speed modem developed for use over the switched telephone network. The second method, referred to as direct sequence spread spectrum (DSSS), has its origin in efforts developed during World War II to overcome enemy jamming.

Orthogonal Frequency Division Multiplexing

OFDM represents one of two "transmission" methods used in BPL field trials. The reason for the quotations surrounding the term transmission is that OFDM actually represents a transmission method under which multiple carriers are used instead of a single carrier. Those multiple carriers are then modulated such that each carrier conveys information.

OFDM dates to the 1980s in modem technology, with the first 9,600-bps modem developed by Telebit Corporation for use over the switched telephone network based on the use of multiple carriers to transmit information over the voice telephone passband. Until the introduction of the Telebit TrailBlazer, modem modulation was based on varying a single carrier. The Telebit TrailBlazer modem was unique for its time because it modulated a series of carriers at predefined frequencies across the telephone passband. During the initiation of a call over the switched network, the originating TrailBlazer modem would generate a series of tones across the telephone passband. The TrailBlazer modem at the opposite end of the communications channel would note which carrier tones were received with sufficient power and a low amount of distortion such that they could be used by the originating modem to modulate data. Then, the receiving modem would convey this information to the originating modem, which would use the "good" carrier tones to modulate data. Thus, OFDM provided a mechanism to bypass frequencies that would adversely affect data modulation.

Overview

OFDM is based on frequency division multiplexing (FDM). FDM represents a technology that enables multiple signals to be simultaneously transmitted over a single transmission link, such as a coaxial cable, a power line, or a wireless system. Under FDM, each signal occurs within a unique frequency range or band. As we noted earlier in this book when we briefly discussed DSL in Chapter 1, that technology is based

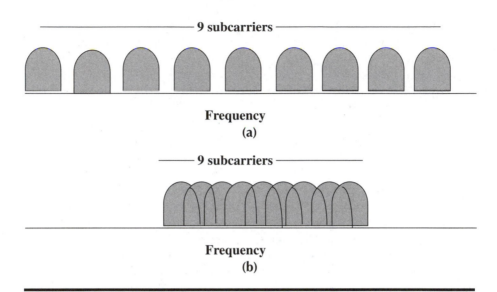

Figure 3.6 Comparing FDM and OFDM. (a) FDM using nine subcarriers. (b) OFDM using nine subcarriers.

on subdividing the telephone twisted-wire pair by frequency into uplink and downlink frequency bands, both of which are located beyond the 300 Hz to 3,300 Hz passband used for voice communications.

Under OFDM a large number of carriers are used. Those carriers are spaced apart at precise frequencies, which provides the "orthogonality" associated with the term and prevents demodulators from seeing frequencies other than their own.

Spectral Efficiency

Although OFDM is similar to FDM, it is much more spectrally efficient. To obtain this spectral efficiency, OFDM spaces subcarriers much closer together until they overlap one another. To visualize the difference between FDM and OFDM, consider Figure 3.6. Figure 3.6a illustrates an FDM system with nine subcarriers, each spread apart by frequency from one another. In comparison, Figure 3.6b illustrates an OFDM system that also consists of nine subcarriers. In comparing the two you will note that the OFDM system's subcarriers not only are closer to one another but, in addition, overlap. The ability for subcarriers to overlap occurs by locating frequencies that are orthogonal. This means that mathematically they are perpendicular, which allows the spectrum

of each subchannel to overlap another without causing channel interference. As you can note by comparing Figure 3.6b with Figure 3.6a,
OFDM also reduces the amount of bandwidth required to transport
information.

The overlapping of signals under OFDM presents a demodulation
challenge. To demodulate the signal requires a discrete Fourier transform (DFT). Fortunately, many chip makers now offer fast Fourier
transform (FFT) chips, the use of which within modems facilitates the
demodulation process.

Because OFDM is based on multiple carriers, it is sometimes referred
to as multicarrier or discrete multitone modulation. It is the modulation
technique used for digital television in Europe, Japan, and Australia
as well as in the IEEE 802.11 series of high-speed wireless LANs and
forms the basis for the Asymmetrical DSL (ADSL) standard.

Modulation

Modulation represents the process whereby a carrier wave operating
at a particular frequency is modified to impress information. There are
three basic kinds of modulation — amplitude, frequency, and phase
shifting. Under amplitude modulation, the carrier wave is modulated
in step with the digital message, which consists of a stream of 1's and
0's. That is, one amplitude is used to represent a binary "1" and a
second amplitude is used to represent a binary "0." Because the
amplitude shifts between two values, this technique is also referred to
as amplitude shift keying (ASK).

Frequency modulation represents the first method used by modem
designers to modulate information. Under frequency modulation, one
frequency is used to represent a binary "1" whereas a second frequency
is used to represent a binary "0." Because frequencies shift from one
to another based on the composition of the bits in a message, this
modulation technique is also referred to as frequency shift keying
(FSK).

A third modulation method results in the phase of the carrier wave
being shifted based on the content of the message stream. Because a
carrier wave can be in phase or out of phase, based on varying the
phase angle, it becomes possible to represent multiple bit values with
each phase change. For example, consider a phase modulation method
that varies phase by 0°, 90°, 180°, and 270°; with four phase changes,
it becomes possible to represent two bits per phase change, as indicated in Table 3.2. This technique, in which two bits are assigned to

Table 3.2 Dibit Encoding

Phase	Bits encoded
0	00
90	01
180	10
270	11

each phase change, is referred to as dibit encoding. As you might expect, a technique in which three bits are assigned to each phase change is referred to as tribit encoding.

When dibit encoding is used, the modulation rate is one-half the data rate and 2^2, or 4, phase angles are required to represent all bit combinations. Similarly, if we pack three bits per phase change we would require 2^3, or 8, phase changes and the modulation rate would be one-third the bit rate.

If you think of the possible phase angles as representing a pie, as the number of phases used to represent groups of bits increases, the size of the slice of the pie decreases. This means that it becomes harder to discriminate small phase changes from large phase changes and places an upper limit on the number of phase changes that can be used to modulate data prior to slight variances in line noise, resulting in a relatively high error rate.

Combined Modulation Methods

One of the more interesting types of modulation overcomes some of the problems associated with too many phase angle changes by combining amplitude modulation with phase modulation. The resulting modulation method, which is referred to as quadrature amplitude modulation (QAM) or amplitude–phase keying (APK), results in the ability to encode many bits into each signal change that represents both an amplitude and a phase change.

OFDM Implementations

There are several methods commonly employed to implement OFDM. Most methods commence with the use of an error correction scheme to protect the data elements from corruption or recognize the occurrence of one or more errors within a packet. To protect data elements from corruption, the data stream is operated on using a forward error correcting code. In comparison, if the error correction method is designed simply to retransmit packets when one or more bit errors are detected, data is packetized and a check algorithm is used to add a check character to the end of the data block. At the receiver, the same algorithm is applied to the data in the packet. Then, the locally computed checksum is compared to the received checksum. If the two match, the packet is assumed to be received without any errors. If the two checksums do not match, one or more bit errors are assumed to have occurred and the receiver will transmit a negative acknowledgment that informs the originator to retransmit the packet. If data to be transmitted needs protection, then an encryption algorithm is applied to the data stream. In the wonderful world of BPL field trials, because the power line represents a LAN, it is possible for a subscriber to "read" packets addressed to other subscribers. Thus, some field trials involve the use of data encryption to make intentional reading of other subscriber packets meaningless.

Once the data stream is encoded and packetized, it is modulated. Over the past decade several versions of QAM were used with OFDM. To provide readers with an indication of the various QAM modulation methods in use, we can turn to the IEEE 802.11a standard, which defines the operation of wireless LANs in the 5-GHz frequency band. Under that standard, four modulation methods are defined. As indicated in Table 3.3, the IEEE 802.11a version of OFDM defines the use of

Table 3.3   IEEE 802.11a OFDM Modulation Support

Data Rate (Mbps)	Modulation Method	Coding Rate	Coded Bits/ Subcarrier	Coded Bits/OFDM Symbol	Data Bits/OFDM Symbol
6	BPSK	½	1	48	24
9	BPSK	¾	1	48	36
12	QPSK	½	2	6	48
18	QPSK	¾	2	96	72
24	16-QAM	½	4	192	96
36	16-QAM	½	4	192	144
48	16-QAM	¾	4	288	192
54	64-QAM	⅔	6	288	216

binary phase shift keying (BPSK), quadrature phase shift keying (QPSK), and two versions of QAM. Depending on the modulation method, coding rate, and number of coded bits per subcarrier, the IEEE 802.11a versions of OFDM can support data rates ranging from 6 Mbps to 54 Mbps.

From Table 3.3, the IEEE 802.11a OFDM modulation method involves the division of a serial data stream into groups of one, two, four, or six bits, with the number of bits based on the data rate. For example, if a data rate of 48 Mbps is selected, each set of four bits in the serial data stream is mapped into a 16-QAM constellation.

Multiple Access Capability

One of the additional advantages associated with the use of OFDM is the fact that it allows subcarriers to be assigned to different users. For example, subcarriers 1, 3, 5, and 7 could be assigned to user A, and subcarriers 2, 4, 6, and 8 could be assigned to user B. Thus, OFDM can also be considered to represent a multiple access technique because an individual tone or group of tones can be assigned to different users. Users can be assigned to a predefined number of tones when they have information to be transmitted or they can be assigned to a variable number of tones based on the amount of information to be transmitted. The assignment of tones is controlled at the Media Access Control (MAC) layer, which uses a scheduling algorithm to assign resources based on user demands.

Direct Sequence Spread Spectrum

Direct sequence spread spectrum (DSSS) represents one of two transmission methods the origins of which date to the early 1940s and which were developed to overcome the effects of enemy jamming. Both DSSS and frequency hopping spread spectrum (FHSS) can be traced to the Hollywood actress Hedy Lamarr and pianist George Antheil, who described the use of a spread spectrum communications method to control torpedoes in 1941 and subsequently received a U.S. patent for it. Unfortunately, spread spectrum communications were not taken seriously by the U.S. government and it was essentially glossed over until the 1980s when its considerable advantages were recognized. Since then, spread spectrum communications have been adopted for use in wireless LANs, satellite positioning systems, Bluetooth, and other applications. Whereas FHSS has been adopted for low-speed communications, DSSS has found considerable use in higher speed communications, such as the IEEE 802.11a, 802.11b, and 802.11g wireless LAN transmission methods.

Overview

Spread spectrum communications to include DSSS represents a transmission method, not a modulation technique. That is, using DSSS, you can transmit a signal using FSK, PSK, or another modulation method.

DSSS combines each data signal at the transmitting station with a higher data rate bit sequence. That sequence is referred to as the chipping code and functions as a redundant bit pattern for each bit that is transmitted. Thus, instead of transmitting a single data bit at a time, a series of redundant bits is transmitted. Then, if one or more bits in the bit pattern is damaged during transmission, the original composition of the bit can be recovered due to the redundancy of the bit's transmission. In fact, a simple majority rule can be used. Thus, if the data bit is a binary 0 and a chipping code of 5 results in the transmission of five binary 0's, which are received due to noise as 10010, because there are three binary 0's and two binary 1's, the receiver would assume that a binary 0 was transmitted.

Justification for Use

Because DSSS transmits a signal spread over a much wider frequency than the minimum bandwidth required for transmitting only data bits, an obvious question is if this spreading is justified. We can commence our investigation using Shannon's law, which describes channel capacity in terms of the signal-to-noise ratio (SNR) on a channel. That is,

$$C = W \log_2 (1 + S/N)$$

where
C = channel capacity in bps
W = bandwidth in Hz
S = signal power
N = noise power

From the above equation we can note that by increasing the bandwidth (W) we can decrease the SNR without decreasing the channel capacity (C). Thus, the process gain (G), which represents the

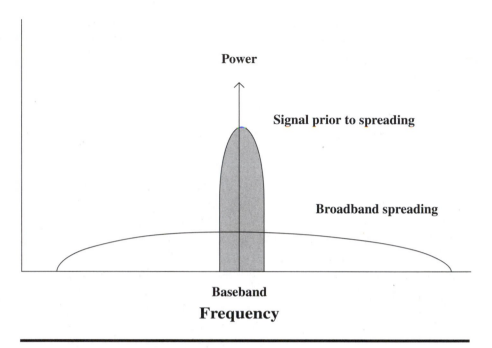

Figure 3.7   Comparing a baseband signal to its broadband spreading.

performance increase for a wideband system and is described mathematically as

$$G = B/I$$

where
B = radio frequency (RF) bandwidth in Hz
I = information rate in bps

shows that performance can be increased without requiring a high SNR.

Under spread spectrum communications, the baseband signal, which represents the original data modulated at a particular frequency, is spread out over a broad frequency range. Figure 3.7 illustrates the relationship between a baseband signal and its spreading over a range of frequencies.

Once a signal is spread over a range of frequencies, the receiver despreads the received signal. In doing so, the receiver uses the same spreading algorithm to recover the data.

Advantages

From Figure 3.7 it's obvious that a spreading technique requires the use of a wider frequency band than when a baseband signal is directly modulated. Fortunately, the requirement for more frequency is compensated for by a series of advantages associated with spread spectrum communications. Those advantages include resistance to interference, resistance to fading, and resistance to jamming.

Because spread spectrum uses a spreading algorithm, transmission is spread over a range of frequencies, which minimizes any potential interference, because interference normally occurs within a close range of frequencies. Concerning fading, wireless transmission can reflect off various objects, which can result in multiple paths by which a signal propagates toward a receiver. Those reflected paths can interfere with the direct transmission path, resulting in a phenomenon called fading. Because a spread spectrum receiver uses the same spreading algorithm as the transmitter, it becomes possible to accept the direct path and reject reflected signals, thus providing resistance to fading.

A third major advantage of spread spectrum concerns its resistance to jamming. Because the transmitted signal is spread over a range of frequencies, it becomes harder to jam than a signal transmitted on a fixed frequency. This resistance to jamming is more pronounced when the military uses spread spectrum because spreading by frequency occurs based on the use of a pseudorandom number generator. When used in a commercial application, the need for interoperability means that all equipment must use the same method of frequency spreading.

Operation

The DSSS process is based on the generation of a sequence of pseudorandom numbers (PRNs). The PRN is subdivided into groups of bits that are referred to as the chipping code, which is applied to each data bit. This application normally occurs via the use of modulo-2 addition of each chipping code bit with each data bit. Once the single data bit is spread into a sequence of bits, those are then modulated using one of several modulation methods, such as FSK or PSK. Then, a mixer is used to multiply the RF carrier and the result of XORing the data bit by the PRN bit sequence. This process causes the concentrated RF signal to be replaced with a wideband signal.

To illustrate the operation of DSSS communications, let's assume that a 5-bit chipping code is applied to each information bit. Under DSSS, each information bit is modulo-2 added to each bit in the chipping code and then transmitted. Table 3.4 illustrates the transmission of the information bits 11100 based on a chipping code of 10101.

Table 3.4   DSSS Coding Example

Information bits	11100
Chipping code	10101
Transmitted bits (modulo-2 addition)	
For first information bit:	10101
For second information bit:	10101
For third information bit:	01010
For fourth information bit:	01010
For fifth information bit:	01010

Note that because a 5-bit chipping code is used, the five information bits are transmitted as a sequence of five 5-bit groups, or 25 bits.

In examining the entries in Table 3.4, note that the chipping code is modulo-2 added to each information bit. That is, the first information bit located at the extreme right (0) is modulo-2 added to each bit in the chipping bit code (10101). This modulo-2 addition process results in the generation of the bit sequence 10101. That bit sequence is then transmitted to represent the information bit of binary 0. Next, the second information bit (0) is modulo-2 added to each chipping code bit (10101) to generate the bit sequence 10101, which is transmitted to represent the second information bit, the value of which is also binary 0. This process continues, with each information bit being modulo-2 added to the chipping code bits to spread each bit.

At the receiver, the first information bit arrives as the binary sequence 10101. Because a modulo-2 subtraction process occurs using the same chipping code, the received binary sequence is modulo-2 subtracted using the chipping code 10101. Doing so results in the generation of five binary 0's, which the receiver, using a majority rule process, accepts as a received binary 0. Next, the second information bit arrives as an encoded group of five bits, as indicated in Table 3.4 as 10101. The result of performing modulo-2 subtraction on these five bits using the chipping code of 10101 is again a sequence of five 0 bits, on which the receiver uses the majority rule to denote as a received binary 0. For the third information bit, which is a binary 1, the modulo-2 addition process spreads the bit into the 5-bit sequence 01010. At the receiver, the received sequence (01010) is modulo-2 subtracted using the chipping code 10101. This modulo-2 subtraction process results in the generation of the five bits 11111. Once again, using the majority rule, the receiver assumes that the third bit received was a binary 1.

Because the chipping code adds redundancy to the information being transmitted, this permits a receiver to recover the original data if one or more bits is damaged or corrupted during transmission. Of course, the ability to recover from transmission impairments depends on the length of the chipping code used and the duration of each transmission impairment. However, when data can be recovered, it is done so without the necessity to have the origination retransmit. Now that we have an appreciation for DSSS, we can turn our attention to a version of that transmission method that has proved popular in cell phones and that may gain equal popularity when used in BPL systems. That transmission method is code division multiple access (CDMA), a digital cellular technology that uses spread spectrum techniques.

Code Division Multiple Access

CDMA represents a transmission method where individual data streams are encoded with PRNs and transmitted over the full spectrum allocated for this digital cellular technology. CDMA was originally known as IS-95 and competes with GSM (Global System for Mobile) technology for dominance in cellular communications. The original version of CDMA is now referred to as cdmaOne, and the wideband version of CDMA forms the basis for the evolving third-generation (3G) network. In between, there are several variations of CDMA. In examining CDMA, we will focus our attention on how it operates in comparison to frequency division and time division systems and the advantages associated with its use.

Overview

In an RF environment, two resources can be used to separate transmissions from individual users — frequency and time. When frequency division is used, each communicator is assigned to a particular frequency. This method of communications is referred to as frequency division multiple access (FDMA) and is illustrated in the left portion of Figure 3.8. The second RF resource, time, can be subdivided into slots that are allocated to different communicators. This type of system is referred to as time division multiple access (TDMA) and is shown in the middle portion of Figure 3.8. In comparison to FDMA and TDMA, CDMA uses unique spreading to spread the signal over the entire

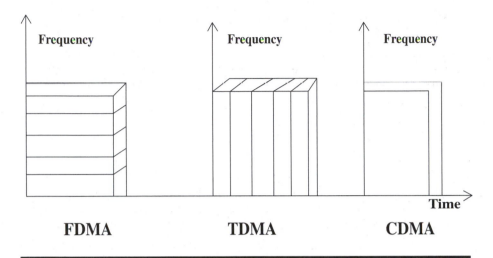

Figure 3.8   Comparing access methods.

spectrum. Those codes spread the baseband signal prior to transmission, resulting in the spreading code used to identify connections. The right portion of Figure 3.8 illustrates the CDMA use of frequency and time.

Coding

The spreading codes used by CDMA represent a carefully designed binary sequence of 1's and 0's that create a uniqueness to identify each particular call. Similar to DSSS, the rate of the spreading code is referred to as the chip rate or chipping code. At the receiver, a correlator is employed to despread the wanted signal, which is then passed through a narrow bandpass filter. Unwanted signals are neither despread nor passed through the filter.

Advantages

The use of CDMA allows for the previously noted advantages associated with DSSS. In addition, it becomes possible for more users to share a given range of frequencies, an important consideration both for cell phone operators as well as for a BPL system in which a large number of potential customers within a small geographic area could be supported by the use of a CDMA cell. At the time this book was written, CDM was in limited use in a European version of BPL referred to as power line communications (PLC). A single-carrier version of CDMA was being used to provide a data transmission rate of approximately 1 Mbps. In the use of CDMA in Europe, it occurred as a transmission method over the power line instead of as a wireless technology. Currently, a number of different proposals are being considered for a multicarrier version of CDMA that, when implemented, could result in higher data rates. In fact, engineers are working on developing a version of CDMA based on OFDM; however, the technology may not be available until the end of this decade.

Network Access

We can view both the electric utility company power line and the internal electric wiring in a home or office as representing a network. Thus, we need a network control protocol that regulates traffic from terminal devices onto the electric wiring that forms the network. In regulating traffic, the network protocol should prevent multiple devices from communicating at the same time, because this action would result in the collision of data. Because there is no limit to the distance between transmitting devices, the network protocol used by Ethernet LANs, Carrier Sense Multiple Access with Collision Detection (CSMA/CD), cannot be used. Instead, the CSMA/CA network protocol widely used in IEEE 802.11 wireless LANs as well as by Apple's Local Talk wired LAN was selected by many electric utility company equipment vendors as their network protocol. Thus, in this section we will turn our attention to the operation of the CSMA/CA network access protocol.

CSMA/CA Overview

The Carrier Sense Multiple Access with Collision Avoidance (CSMA/CA) network protocol represents a layer 2 protocol that controls access to a network. Under CSMA/CA, collision avoidance is used to improve the performance of the network by attempting to reserve its use at a particular point in time to a single device. To do so, it employs a carrier-sensing scheme in which the modulation of data on the network indicates that the network is in use. A device that has data to transmit would first listen to the network. If no carrier is sensed, this would indicate that there is no activity on the network. Then, the device with data to transmit would transmit a jam signal to indicate to other devices that it intends to transmit. After waiting a sufficient period of time for all other devices on the network to receive the jam signal, the device with data to transmit would then begin its transmission. While transmitting, if the device detects the presence of a jam signal issued by another device, it will cease transmission for a random period of time and then repeat the previously described process.

Each device on a CSMA/CA network is technically referred to as a station. Ideally, the overhead introduced by the collision avoidance scheme should be as small as possible while keeping the number of collisions to a minimum. To accomplish this, the range of the random delay, which is also referred to as the contention window, is varied with the traffic load on the network. For example, in the event a collision occurs, the delay is doubled progressively until a successful transmission occurs. Then, the delay is reset to its initial minimal value. Because the overhead associated with the use of the CSMA/CA network access protocol increases as the number of stations on a network increases, it would be ill advised for a utility to use the protocol to cover its entire infrastructure. Instead, in several electric utility field trials, it appears that the substation provides a convenient delimiter, with homes and businesses served by the substation becoming a subnet that uses the CSMA/CA protocol and that is terminated from other subnets via the use of a layer 3 router. Thus, limiting the potential number of BPL subscribers on a network to the number of homes and offices served by a substation ensures that CSMA/CA network delays are minimized.

Chapter 4

Interference and FCC Action

Electric utility power lines were not designed for data transmission. Instead, as we are well aware, they were constructed to transport power at 50 Hz in Europe and at 60 Hz in North America. This design, which dates back approximately a hundred years, resulted in power lines constructed without considering the need for insulating radio frequency (RF) energy associated with data transmission. Thus, the transmission of data over a power line results in the line turning into an antenna, generating unwanted radio frequency interference (RFI).

In this chapter we will turn our attention to the broadband over power line (BPL) RFI problem, examining why power lines act as antennas and how certain frequencies used for communications are adversely affected by the transmission of data over power lines. Because the Federal Communications Commission (FCC) is responsible for regulating communications in the United States, we will also examine their actions with respect to the transmission of data over power lines to include a recent FCC ruling.

4.1 BPL Interference

The technology behind the use of power lines must operate over facilities designed to move power at 50 Hz or 60 Hz. Those facilities

Table 4.1   Characteristics of a Good Transmission Facility

Low level of transmission loss
Low level of electromagnetic radiation
Immunity to external interference
Constant level of attenuation
Constant level of propagation delay
Low level of noise

are far from providing a good mechanism for the transmission of RF-modulated information. Let's first discuss the characteristics of a good transmission facility. Then, after we have an understanding of the characteristics of a good transmission facility, we can compare and contrast those characteristics to the transmission properties of power lines.

Good Transmission Characteristics

There are several transmission facility characteristics that enable data to be recovered at a receiver with a minimum probability of an error occurring while at the same time minimizing potential interference to other transmissions occurring near the facility. Table 4.1 lists six of the major characteristics of a transmission facility that serve to minimize the error rate and level of interference with other transmission facilities.

Transmission Loss

The first entry in Table 4.1, a low level of transmission loss, governs the need for amplifiers as well as the ability of a signal to be received correctly. When a transmission facility has a low level of transmission loss, this reduces the need for amplifiers, permitting such devices to be installed further from one another. This in turn reduces the cost associated with constructing a transmission facility.

Because the level of transmission loss affects the characteristics of a received signal, a low level of transmission loss results in the reception of a better signal. This in turn reduces the probability of a transmission error. Thus, the level of transmission loss affects both the need for amplifiers and the error rate experienced on a transmission facility.

RF Radiation

A conductor can be viewed as a potential antenna based on its balance and shielding. For example, in Chapter 2 we discussed the Faraday effect, which results in electromagnetic radiation occurring when current flows through a conductor. In that chapter we discussed the right-hand rule, with current flowing in a conductor in the direction of a person's upright thumb, resulting in radiating energy flowing in the direction of a person's other four fingers, which are bent to touch the palm.

RF radiation represents a two-way street. External radiation can induce current into a conductor, which can adversely affect transmission. Similarly, transmission on a conductor can result in the radiation of energy, which can disturb certain types of transmissions within close proximity of the conductor.

There are two primary methods used to minimize the effects of electromagnetic radiation. One method results in shielding a conductor, whereas the second method attempts to balance unshielded conductors, typically by twisting the conductor so that the electromagnetic radiation at different locations on the twisted conductor cancels itself. If you examine certain types of conductors, such as CAT 5 wiring, you will note that these types of conductors have pairs of wires that are twisted. In fact, the CAT 5 specification defines the number of twists such that the wiring provides a good balance in the proximity of other conductors. Thus, both shielding and twisting of conductors can be used to minimize electromagnetic radiation from a conductor as well as provide the conductor with immunity to external interference.

Attenuation

Attenuation represents a reduction of signal strength due to the resistance, inductance, and capacitance of a conductor. A good transmission characteristic of a circuit is a uniform level of attenuation across the frequency spectrum. The reason a uniform level of attenuation is a good characteristic is that it enables an amplifier to boost a signal by a set amount over the range of frequencies used by the signal. Otherwise, an amplifier would need to boost certain portions of a signal differently than other portions of a signal, resulting in a rather burdensome and costly method.

One physics principle states that high frequencies attenuate more rapidly than lower frequency signals. Thus, the amount of attenuation experienced on a circuit varies over the frequency spectrum. This uneven level of attenuation provides a challenge for enhancing data transmission because an amplifier would have to be designed to boost high frequencies more than low frequencies. Because the selective boosting of frequencies would be cost prohibitive, it is common to use an amplitude equalizer prior to using an amplifier. The amplitude equalizer places an inverse mirror image of the signal loss across the frequency spectrum, resulting in a total near-uniform loss occurring across the set of frequencies. Thus, after the signal is equalized, it is then amplified.

Propagation Delay

Signals flowing on a conductor consist of a series of frequencies that experience different delays. For example, low and high frequencies commonly experience more delay than frequencies near the center of the spectrum being used. This propagation delay effect can result in data errors due to some signals inadvertently arriving at a receiver prior to other signals. For example, consider an elementary frequency shift keying (FSK) modulation method, in which a "1" bit is modulated by placing a tone at 1,030 Hz on a transmission line while a "0" bit is modulated by placing a tone at 1,230 Hz on the line. Let's also assume that the baud or signaling rate is 1,000 signals per second, or .001 seconds per signal. Then, if the delay experienced by a tone occurring at 1,030 Hz is just slightly more than 1 ms above the delay experienced by a tone occurring at 1,230 Hz, it becomes possible for a binary 1 to be misinterpreted as a binary 0. When we consider the characteristics of a good transmission facility, its propagation delay should be as near uniform as possible across the frequency spectrum.

Similar to attenuation, variances in propagation delay can be resolved through the use of a delay equalizer. The delay equalizer induces a delay that represents a reverse mirror image of the delay encountered over the frequency band, in effect resulting in a near-uniform delay over the spectrum. Because the delay is now near a uniform level, the delay equalizer prevents the previously mentioned problems whereby one modulated bit can inadvertently arrive at a receiver ahead of a previously modulated bit, resulting in the occurrence of a bit error.

Noise

Another characteristic of a good transmission facility is a low level of noise. To understand why the level of noise is an important consideration, we can turn our attention to the meaning of the signal-to-noise ratio (SNR). If the signal is below the level of noise, a receiver cannot distinguish the signal from the noise, resulting in an inability to recover the transmitted signal. If the signal increases or the level of noise decreases, the signal will eventually be above the level of noise, allowing the receiver to be able to distinguish the signal from the noise. Thus, the level of noise on a conductor plays an important role concerning the ability of a receiver to correctly interpret transmitted signals.

Although the degree that the level of a signal is above that of noise enables certain signals to be received better than other signals, one cannot arbitrarily continue increasing the power level of a signal. The FCC places limits on signal power to prevent signals at certain frequencies from interfering with signals transmitted at other frequencies. After all, you wouldn't want a CB (citizen's band) radio operator talking about a sale at a local department store to interfere with a pilot making his final approach when landing at a nearby airport.

Now that we have an appreciation for the general good characteristics of a transmission facility, let's examine those characteristics when data is transmitted over electrical power lines. In doing so, let's also discuss what is probably the major cause of interference generated by the transmission of data over power lines: the composition of medium- and low-voltage lines. Because the power lines do not have RF shielding, they function as a long antenna that radiates unwanted signals when data is transmitted over them and also picks up unwanted signals.

Power Line Transmission Properties

An electric power line was designed to deliver power at 50 Hz to 60 Hz. Although data can be modulated and transmitted over the same power line at different frequencies, a series of obstacles needs to be overcome to enable data transmission to occur with a relatively low error rate. In this section we will examine the effect of power line properties on the transmission of data, including interference resulting from transmitting data over power lines.

Noise

A variety of factors influence the generation of randomly occurring electrical noise on power lines. First, atmospheric conditions, including solar flaring (sunspots) and lightning, randomly generate noise. Second, when air conditioners, heat pumps, regular household appliances, and other energy-consuming devices are turned on, this results in a surge of current, which generates noise. A third source of noise can include nearby power lines as well as arcing and discharge at locations where there are faulty connectors or dirty insulators. Perhaps the most prominent cause of noise on a power line is the current flowing over the line. As previously mentioned in this book, high-voltage lines are very noisy, which currently precludes their use for data transmission. Although medium- and low-voltage lines have a degree of inherent noise, they are less noisy than high-voltage lines. Because power lines are not RF shielded, the line functions as a long antenna. Thus, power lines will receive over-the-air RF transmission, such as broadcast radio stations as well as other fixed and mobile radio sources. This means that the antenna effect can result in a power line functioning as a receiver, with received signals acting as noise with respect to data transmitted on the line.

Because power lines are not shielded, they normally have considerably more noise than a line specifically used for data transmission. Thus, the primary method to minimize interference caused by various sources of noise upon data transmission over power lines is to select frequencies that minimize interference. Later in this chapter, we will note frequencies selected by the FCC for BPL operations.

Attenuation

As noted earlier in this chapter, attenuation represents a reduction in signal strength that occurs due to the resistance, inductance, and capacitance encountered as a signal flows on a transmission facility. Although transmission distance represents a major cause of attenuation, other factors, including changes in a feeder line from an overhead to an underground facility and spliced and tapped connections, also affect attenuation. In addition, because high frequencies attenuate more rapidly than lower frequencies, the level of attenuation is not uniform.

One way to minimize the effect of attenuation is to reduce transmission distances. Another method is to use amplifiers that boost an attenuated signal. In the United States, the implementation of BPL field trials resulted in data signals traveling less than a mile. This is because the substation typically serves as a delimiter for each BPL subnet and the distance between a substation and the home or office served by the substation is at most approximately 1 mile. Because high frequencies attenuate more than low frequencies over longer distances, the relatively short BPL transmission distance generally alleviates the need for attenuation equalizers, enabling amplifiers to be used at distances between 1,000 ft and 2,500 ft to boost the modulated signal.

Propagation Delay

Similar to our discussion of high-frequency attenuation, experiencing different delays with different frequencies is relatively insignificant when transmission distances are short. Thus, the structure of the BPL infrastructure enables electric utilities to overlook propagation delay, which would be a more serious issue if transmission distances were hundreds or thousands of miles instead of typically under a mile.

The Transformer Effect

The neighborhood or distribution transformer closest to the BPL subscriber represents a major barrier to providing data services. Although low-frequency signals to include 60-Hz alternating current can easily pass through a step-down transformer, high frequencies are either severely attenuated or obstructed, resulting in the transformer appearing as an open circuit with respect to the flow of data modulated at high frequencies.

There are three methods currently used to compensate for the open-circuit effect associated with distribution transformers. First, the transformer can be bypassed through the use of couplers. A second method involves amplifying the signal to a sufficient level so that it can pass through the transformer. The third method is to use a WiFi access point, which eliminates the need to either bypass or go through the transformer.

Both bypassing and amplifying a signal to flow through the neighborhood transformer represent a two-headed sword. That is, although each technique is designed to enable signals carried on medium-voltage power lines to flow over low-voltage lines, they also create a degree of magnetic interference. Although the use of a WiFi access point also results in the generation of RF signals, those signals occur in the 2.4-GHz unlicensed industrial, scientific, and medical (ISM) band. Thus, those signals interfere only with other devices that operate in the ISM band, such as cordless telephones and microwave ovens. However, because IEEE 802.11 WiFi signals are primarily based on the DSSS transmission method, most interference is minimized.

Power Line Problems

Previously we noted that shielded cables can be used to minimize radiation. Unfortunately, power lines are not RF shielded, so the use of shielding does not represent a solution to the power line radiation problem.

Lack of Symmetry

Another method to reduce or minimize RFI is to balance unshielded conductors. We can note this by examining twisted-pair telephone and LAN cable, where pairs of conductors are twisted to minimize RFI. If you look at an overhead power line, you will easily observe that such lines are far from being ideal balanced lines for reducing RFI. This is because low-voltage power lines lack the symmetry of the medium-voltage lines. In fact, some of the branches off the main distribution power line that provide electricity for street lights add further to a lack of symmetry, while the neutral conductor typically grounded via a metal rod inserted into the earth at the base of a telephone pole further reduces power line symmetry. Thus, the neighborhood power lines that provide the transmission path into homes and offices can be expected to produce RFI.

Antenna Effect

Due to power lines behaving similarly to an antenna, the transmission of modulated data can result in a considerable amount of RFI, which diminishes as the distance from the power line increases. Some BPL field tests resulted in RFI occurring at distances up to 75 meters for mobile radios and at distances of up to 150 meters for fixed-location radios. In addition, various elements or structures associated with power lines, such as street lamps, can become RF radiators at the high frequencies used in BPL field trials. Needless to say, unwanted transmissions from BPL field trials resulted in many individuals as well as organizations becoming vocal concerning the potential interference caused by BPL.

ARRL Opposition to BPL

One of the more vocal groups against the spread of BPL is the American Radio Relay League (ARRL), a worldwide organization of amateur radio operators headquartered in Newington, Connecticut. The ARRL noted

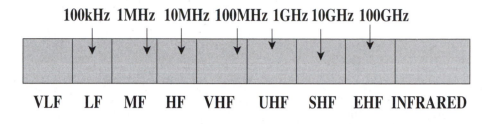

Figure 4.1   The bands in the frequency spectrum.

Table 4.2   Common Communications Applications and Their Frequencies

Application	Frequency
AM radio	535 kHz to 1.7 MHz
Short-wave radio	5.9 MHz to 26.1 MHz
CB radio	26.96 MHz to 27.41 MHz
Television	54 MHz to 88 MHz (for channels 2 through 6)
FM radio	88 MHz to 108 MHz
Television	174 MHz to 220 MHz (for channels 7 through 13)

early on that BPL systems can cause interference to licensed services, including amateur radio transmission. In fact, the ARRL sponsored monitoring of several BPL field trials to determine the level of interference occurring from such trials. During 2005, ARRL documented interference caused by many BPL trials as well as recorded the interference caused by such systems operating between 2 and 80 MHz, which resulted in considerable problems occurring in the amateur radio high-frequency (HF) band. The HF band is between 3 and 30 MHz and is referred to as shortwave frequencies. Figure 4.1 illustrates the frequency spectrum, which includes the frequency bands from very low frequency (VLF) through infrared. To better understand the relationship between BPL frequencies set aside by the FCC, which we will discuss, and other common communications, see Table 4.2. The table indicates frequencies used by six common communications applications.

One of the key problems concerning interference caused by BPL is its operating frequencies that can adversely affect transmission in the HF frequency band. As indicated in Figure 4.1, the HF band, resides between the medium-frequency (MF) and very high-frequency (VHF) bands. The HF band has some special properties not found in other frequency bands. One such property is the ability to communicate at very low power levels around the world without needing repeating devices, such as satellites or towers. This key property is based on the fact that HF signals can bounce off the ionosphere multiple times to reach their destination. Due to BPL adversely affecting the HF frequency band, which is used by tens of thousands of amateur radio operators, the ARRL became one of the leading opponents of BPL. In addition, because BPL interference also adversely affects CB operators, shortwave radio listeners, and other radio users, other organizations and individuals became vocal against the use of BPL. One of the problems mentioned by both groups and individuals was the noise generated from the use of spread spectrum transmission. The use of orthogonal frequency division multiplexing (OFDM) modulation was found to create a series of "radio blips" at approximately every 1 kHz across the radio spectrum, with each blip sounding like a ringing or clicking sound. In fact, the ARRL placed recorded sounds monitored near several BPL field trials on their Web site to indicate the interference caused due to the transmission of data over power lines. The organization has become a proponent for the FCC placing strict limits on BPL to minimize its level of interference.

Interference Prevention

Several methods can be used to either prevent or reduce BPL interference to other RF services. Those methods include minimizing the RF power allowed for BPL transmission, avoidance of locally used frequencies, differential-mode signal injection, the use of filters and signal terminators, and the design of BPL systems such that only one active device occurs in a given area. In this section we will examine the potential of each method to prevent or minimize the effect of BPL interference on other RF services within the general area of power lines.

Power Level

The power level associated with the modulation of carriers over BPL represents the major factor that governs the level of potential BPL interference with other RF services. Simply put, a low level of power results in minimal interference, whereas a high level of power can potentially result in harmful interference. The U.S. Communications Act of 1934 and FCC rules have long required that unlicensed transmitters, such as BPL, must protect licensed services, such as radio and television, from interference. In addition, unlicensed transmitters must accept any interference to their operations that results from the normal operation of licensed communications services. Under FCC Part 15 regulation, 15.15(c) states "... the limits specified in this part will not prevent harmful interference under all circumstances. Because the operators of Part 15 devices are required to cease operations should harmful interference occur to authorized users of the radio frequency spectrum, the parties responsible for equipment compliance are encouraged to employ the minimum field strength necessary for communications ...." Thus, BPL systems must cease operations if they cause harmful interference to other RF systems.

The design of a BPL system represents a tradeoff with respect to its power level. If too little power is used, data throughput may be significantly reduced, whereas a higher power level that enhances throughput may also increase the potential of interference with other RF systems. One way to accommodate FCC 15.15(c) is to increase the use of repeaters, which enables BPL systems to use a lower power level.

Avoidance of Locally Used Frequencies

A second method that can minimize the potential of BPL interference on other RF systems is to avoid certain frequencies and frequency bands used by other RF systems. As we will note, the FCC issued a list of excluded frequencies that cannot be used by BPL. BPL systems can use modulation techniques that avoid excluded frequencies as well as filtering to minimize the potential of harmonics from causing interference.

Differential-Mode Signal Injection

As we noted earlier in this chapter, one method that can be used to minimize interference is to have a balanced transmission system. Although power lines are not balanced, it is theoretically possible to inject signals of equal magnitude and opposite phase that are simultaneously transmitted on a pair of lines. This action in effect would result in the cancellation of radiation due to transmission, causing a balance in the lines. Although not required by FCC rulings, it is possible for BPL manufacturers to develop differential-mode signal injection products that could reduce signal radiation, which in turn would reduce potential interference with other RF systems.

Using One Active Device per Area

During BPL field trials it was noted that BPL devices located within a given area tend to transmit one at a time. When this situation occurs, BPL signals do not aggregate and therefore produce less potential interference. Thus, if BPL operations were relegated to using one power line phase in a given area and only one signal injection point per wire, this configuration would minimize potential interference.

Selection of Signal Carrier

One additional technique that could be used to minimize potential BPL interference involves the applicable selection of a signal carrier. Doing so may make it possible to locate one or more frequency bands at which BPL signal radiation is relatively low, reducing potential RFI. Unfortunately, this method of interference minimization requires making detailed frequency measurements to identify where certain frequencies should be avoided. In addition to selecting an applicable signal carrier, this technique also requires careful consideration of coupler placements. Thus, although this method could permit BPL devices to operate at higher signal power levels while minimizing potential interference, its significant implementation requirements have precluded its use in field trials.

4.2 The Regulatory Effort

At the federal level, one of the goals of the Telecommunications Act of 1996 was to promote universal access to broadband services. To accomplish this goal, potential subscribers to broadband services must be offered the service and it must also be affordable. Because DSL (Digital Subscriber Line) requires subscribers to be within 18,000 ft of a telephone company central office, and cable service is available to only a little over half of the homes in the United States, BPL can be viewed as a viable alternative to meet the goals of the previously mentioned legislation.

Role of the FCC

The FCC is tasked with the regulation of the communications spectrum. Its regulatory task includes the assignment of frequency usage by various RF services such that one service does not interfere with transmission on another service. Although BPL represents a wired communications system, because its transmission over power lines results in the generation of interference that can adversely affect other licensed services, the FCC established so-called "excluded frequency bands." As we will note later in this section, the excluded frequency bands represent frequencies where BPL cannot operate due to the potential interference to other licensed services.

The regulatory history associated with BPL in the United States spans less than three years, a relatively short period in the history of communications regulation. The regulatory process commenced with the release of a notice of inquiry (NOI) in April 2003, ET 03-104 titled, "Inquiry Regarding Carrier Current Systems, including Broadband over Power Line Systems." The purpose of the NOI was to obtain comments from the public concerning how existing FCC rules should be revised to promote the deployment of BPL systems while protecting existing licensed services. The NOI was followed by a notice of proposed rulemaking, ET 04-37 titled, "Carrier Current Systems, including Broadband over Power Line Systems: Amendment of Part 15 regarding New Requirements and Measurement Guidelines for Access Broadband over Power Line Systems." The proposed rule was intended to reduce regulatory uncertainty by BPL adopters, thereby facilitating the use of the technology while protecting other FCC licenses from harmful interference caused by BPL operations. A final rule was issued in October 2004. That ruling was in the form of the adoption of a report and order. As we will note, the final rule, although supporting BPL as a "third wire" for broadband service, did not make everyone happy. In fact, the number of comments received by the FCC exceeded 6000, which is probably a record. Another interesting fact was that a significant majority of comments were against the deployment of BPL. However, in spite of the potential interference to licensed services, the FCC final rule gave BPL a significant boost and can be viewed as a green light to utilities sitting on the sidelines waiting to see if the interference problem will be a significant barrier to the rollout of services.

FCC Action

Because of the interference caused by data transmission over power lines, the FCC studied the problems associated with BPL transmission. During the latter part of 2004, the FCC approved a series of rules designed to limit interference from the transmission of data over power lines to other RF devices, such as amateur radio and aircraft receiver operations. Those rules require BPL operators to avoid operating in certain excluded frequency bands as well as avoid operating on certain frequencies near sensitive operations, such as Coast Guard or radio astronomy stations. BPL operators must also employ devices that can switch frequencies of operation if they cause interference, and such devices must be capable of being shut down remotely. In addition, a publicly available BPL database must be created to help identify and resolve harmful interference, and BPL operators must consult with public safety agencies, federal government organizations, and aeronautical stations.

Frequency Use

The FCC rules issued during 2004 include a list of frequency bands that BPL systems cannot use due to potential interference with other previously licensed radio usage. One can invert the list of exclusions to obtain a list of FCC-authorized BPL transmitting bands of operation. Table 4.3 provides a list of FCC-authorized BPL transmission bands derived by this author based on the list of excluded frequency bands. Table 4.4 presents the excluded frequency bands listed by the FCC.

Table 4.3 FCC-Authorized BPL Transmission Bands (in MHz)
1.705 to 2.850
3.025 to 3.400
3.500 to 4.650
4.700 to 5.450
5.680 to 6.525
6.685 to 8.815
8.965 to 10.005
10.100 to 11.275
11.400 to 13.260
13.360 to 17.900
17.970 to 21.924
22.000 to 74.800
75.200 to 80.000

Table 4.4   FCC BPL-Excluded Frequency Bands
2,850–3,025 kHz
3,400–3,500 kHz
4,650–4,700 kHz
5,450–5,680 kHz
6,525–6,685 kHz
8,815–8,965 kHz
10,005–10,100 kHz
11,275–11,400 kHz
13,260–13, 360 kHz
17,900–17,970 kHz
21,924–22,000 kHz
74.8–75.2 MHz

The assignment of excluded frequency bands in effect informed utilities that they could use inverted frequencies from the prohibited list. Although most industry pundits viewed the FCC ruling as a positive development for the wide-scale deployment of BPL, amateur radio operators were anything but pleased with the establishment of excluded frequency bands. One of the key reasons for the FCC rules being regarded as detrimental to amateur radio service is that an inversion of excluded frequency bands results in many authorized bands overlapping amateur radio operations. To appreciate the ire of many amateur radio operators toward the FCC ruling, consider Table 4.5, which indicates the overlap of authorized BPL frequency bands with amateur radio bands. As indicated in the table, there are eight frequency bands BPL can operate on that could cause potential interference to different amateur radio operations.

Because of the overlap between authorized BPL frequency bands and amateur radio bands, several amateur radio operators have set up Web pages to assist other operators in filing complaints about BPL interference. According to some amateur radio operators, it is possible for radio stations operating in the 10-meter and 12-meter amateur bands to hear BPL interference at distances up to 2 miles from overhead BPL transmissions. Radio operators are asked to file complaints via e-mail indicating the frequencies or range of frequencies on which the BPL signal was heard, the level of interference, and a description of the amateur station and its antenna system. Such complaints are being

Table 4.5 Overlap of Authorized BPL Transmission Bands and Amateur Radio Service

Band	Amateur Radio Service
1.705 to 2.850	160 meter
3.500 to 4.650	80 meter
4.700 to 5.450	60 meter
6.685 to 8.815	40 meter
10.100 to 11.275	30 meter
13.360 to 17.900	20 meter
17.920 to 21.924	17 meter, 15 meter
22.000 to 74.800	12 meter, 10 meter, 6 meter

compiled and sent to the BPL operator, the FCC, and the ARRL. This complaint effort, which can be thought of as representing a grassroots lobbying campaign, is aimed at getting the attention of the FCC in hopes that it will further exclude frequency bands for BPL usage that overlap amateur radio bands. Although it is too early to judge the ultimate effect of this lobbying effort, it is quite possible that the FCC could issue additional rules that further restrict the frequencies BPL can use.

Other Regulatory Issues

In addition to the previously mentioned FCC ruling that denoted frequencies BPL systems must avoid, several regulatory issues remain to be resolved. One of the key issues is the classification of BPL services. That is, will BPL be considered by the FCC to represent an information service or a telecommunications service? If the FCC considers BPL to represent a telecommunications service, this means that BPL operations will be subject to regulations defined by the Communications Act of 1934, which means BPL will be viewed as a common communications carrier. Because the FCC previously ruled that cable modem operations are considered to represent an information service, there is a high degree of probability that BPL systems will be ruled similarly. If BPL is considered to represent an information service, it will not be subject to most, if not all, of the regulations applicable to common carriers, including contributing to the universal service fund (USF). The USF, which is added to telephone bills to subsidize telephone service for rural and low-income customers, can make BPL more cost effective if operators do not have to add it to customer bills. Other issues that remain to be resolved include whether Voice-over-IP (VoIP) offered via BPL will be regulated and if such services will have to comply with 911 requirements for emergency communications. In all probability, we can expect a series of rulings from the FCC over the next few years that will resolve these issues.

Chapter 5

BPL in the Home and Office

No discussion of broadband over power line (BPL) transmission would be complete without an examination of how residences and offices can communicate with the BPL infrastructure created by an electric utility. Because electric utilities offering BPL use two methods to cover the last few hundred feet from medium-voltage lines into homes and offices, this chapter will focus on those two methods. As a refresher, one method is to use the low-voltage line routed into the home or office for transmission, and the other method involves the use of wireless over-the-air transmission. The first method requires the use of a power line modem within the home or office to provide compatibility with the BPL equipment used by the electric utility. Because the majority of field trials conducted in North America use BPL equipment at the low-voltage line that is compatible with the HomePlug standard, we will discuss the existing standard as well as an evolving higher speed standard in the first portion of this chapter. In the second part of this chapter, we will discuss the key IEEE 802.11 series extensions, to include the basic 802.11 specification and the 802.11a, 802.11b, and 802.11g extensions as well as the emerging 802.11n extension. This will increase our familiarity with wireless LAN (local area network) technology used by BPL providers as well as make us aware of emerging technology that can be expected to be used to facilitate the bypassing of low-voltage lines in the near future.

5.1 The HomePlug Standard

Readers who have followed the powerline industry may remember a technology called PassPort. A company named Intelogis was responsible for developing PassPort technology, which was based on the use of frequency shift keying (FSK) to send data over electrical wiring. FSK is a relatively simple technology under which a digital "1" is modulated by transmitting a tone at one frequency and a digital "0" is modulated by transmitting a tone at a second frequency. Although it is inexpensive and supports transmission over electrical wiring, it is limited to supporting data rates up to 350 kbps and should not be confused with the technology developed by a company named Intellon, whose transmission scheme is based on the use of orthogonal frequency division multiplexing (OFDM), which was selected by the HomePlug alliance as the standard for powerline networking. The HomePlug standard was developed as an in-home network technology that would allow for existing electrical circuits to provide the transmission medium for interconnecting electronic devices. As we will note, the basic HomePlug specification dates to 2000; work on a new specification can be expected to boost the 14-Mbps data rate of the earlier specification to 200 Mbps.

Evolution

The HomePlug standard dates to March 2000, when the HomePlug Powerline Alliance was founded by 13 industry companies and work commenced on selecting a baseline technology and developing specifications for transmission of data over power lines. This effort resulted in the release of the HomePlug 1.0 specification in June 2001. This specification defined the technology and operation required to transmit data at 14 Mbps over the electrical wiring found in homes and small offices.

The release of the HomePlug 1.0 specification was followed in February 2003 with work on developing a higher speed version of the technology that would enable data rates up to 200 Mbps. That effort is referred to as HomePlug AV. At the time this book was prepared, the HomePlug AV specification was nearing completion.

A third specification being worked on by the HomePlug alliance is the HomePlug BPL specification. This specification denotes the technical and operational characteristics of equipment to be used by service providers and utility companies on low-voltage lines to obtain compatibility with HomePlug equipment, with the latter selected as a

Table 5.1   HomePlug-Certified Products

Bridges
Routers
Adapters (powerline to USB)
Adapters (powerline to Ethernet)
Wireless access point
Security cameras
PCs with HomePlug built-in VoIP (Voice-over-IP) phones
Audio endpoints and speakers

baseline. Because HomePlug 1.0, HomePlug AV, and HomePlug BPL are compatible and interoperable technologies, users with equipment that is compatible with the HomePlug 1.0 specification can communicate with equipment that will shortly reach the market that adheres to the HomePlug AV and HomePlug BPL specifications.

HomePlug Products

Since the initial HomePlug 1.0 specification was defined in 2001, millions of HomePlug products have been shipped worldwide. Each product that complies with the HomePlug specification is considered to represent a certified product. Table 5.1 lists the common categories of HomePlug-certified products that are available for purchase.

Applications

Because homes and offices have electrical outlets throughout the facility, it can be very economical to use the existing electrical wiring within a building for data transmission instead of recabling the telephone line or coaxial cable connection. Recognizing this situation, approximately 30 vendors now manufacture HomePlug-compatible equipment that can facilitate networking within the home and small office. For example, Bell South offers a DSL (Digital Subscriber Line) modem installation service with powerline adapters that can be used to extend Internet connectivity via communications from a DSL modem to other areas within a building. Similarly, Cox Communications provides remote room home networking by using Linksys (recently acquired by Cisco Systems) powerline Ethernet adapters. Other service providers (including AOL Broadband, Comcast, and EarthLink, to name but a few) use powerline Ethernet or powerline USB adapters from various manufacturers to extend networking from a central location in the home or small office to other rooms in a building.

In addition to their use by service providers, powerline adapters are available for purchase from vendor Web sites as well as from a large number of electronics stores, distributors, and national retail outlets, such as Best Buy, Radio Shack, and Wal-Mart. By purchasing such equipment, the homeowner or small business proprietor expands his or her communications network at minimal cost. In fact, I used two powerline Ethernet adapters to extend communications from an access point located in the kitchen of my home to an office located on the second floor over a garage area, where it was difficult to use wireless communications. To extend communications, I simply plugged a powerline Ethernet adapter into an electrical outlet in my office and connected the adapter to the Ethernet port built into the computer in my office. Then, a second powerline Ethernet adapter built into a wireless LAN router/access point was plugged into an electrical outlet downstairs, with the access point cabled to a cable modem that provides access to the Internet. Because the access point was connected to a cable modem, this configuration enabled my wife to use a laptop downstairs to obtain over-the-air connectivity while I worked in my home office upstairs and gained Internet access by communicating with the access point via the electrical wiring in my home. Now that we have an appreciation for the evolution of the HomePlug specifications and some of its applications, let's turn our attention to its underlying technology. In doing so we will discuss the modulation method, frame formats, and the channel access mechanism defined by the HomePlug 1.0 specification.

The HomePlug 1.0 Specification

Similar to other communications specifications, the HomePlug 1.0 specification defines how communications flow over the medium used, to enable different vendors to manufacture devices that will be compatible with one another. To ensure compatibility, the HomePlug 1.0 specification defines the physical layer operation to include the use of OFDM, the Media Access Control (MAC) layer protocol to include frame formats, and such additional features as priority classes or quality of service (QoS), forward error correction, and the use of DES (Data Encryption Standard) for privacy.

Physical Layer

Under the HomePlug 1.0 specification, OFDM is defined as the basic transmission technique used by compatible devices. As a review, under OFDM the available frequency spectrum is subdivided into many narrowband, low-data-rate subcarriers, with each subcarrier perpendicular or orthogonal to the other subcarriers. Because the frequency responses of the subcarriers are overlapping, the result is a high level of spectral efficiency.

Each of the subcarriers used for transmission of data can be modulated through the use of various techniques. The version of OFDM specified for use by the HomePlug 1.0 specification uses 84 equally spaced subcarriers in the frequency band from 4.5 MHz to 21 MHz. Under OFDM, packets of data are simultaneously transmitted over multiple carrier frequencies, providing increased transmission and reliability. Concerning the latter, if noise or a surge in power usage disrupts one or more of the carrier frequencies, the disruption is detected and data transmission is switched to another "good" carrier. This technique is referred to as a rate-adaptive design because the data rate varies in tandem with the number of good carriers in use. Modulation methods specified by the HomePlug 1.0 specification include versions of differential binary phase shift keying (DBPSK) and differential quadrature phase shift keying (DQPSK). Here the term "differential" references the fact that data is encoded as the difference in phase between the present and the previous symbol in time on the same subcarrier. This use of differential modulation enhances performance in environments where rapid changes in phase can occur.

Figure 5.1 illustrates the use of frequency by the HomePlug specification in comparison to the use of frequency by voice telephone and DSL services. As indicated in Figure 5.1, the use of the 4.5 MHz to 21 MHz frequency band minimizes interference with voice telephone and DSL communications that can occur on telephone wires routed near electrical wiring initially installed to deliver power at 60 Hz.

Although DSL also uses OFDM, HomePlug uses this modulation method in a burst mode for full mesh communications. In comparison, DSL uses OFDM for point-to-point communications between the user and the telephone company central office.

Tables 5.2 and 5.3 indicate the encoding techniques used by DBPSK and DQPSK modulation methods, respectively. As indicated in Table 5.2, each input is modulated by either leaving the phase of a carrier unchanged ("0" bit input) or changing the phase of the carrier by 180°

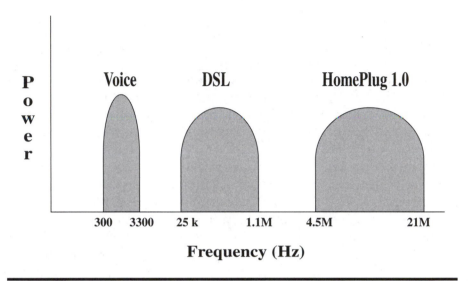

Figure 5.1   Comparing frequency use between telephone wiring and the HomePlug 1.0 specification.

Table 5.2   DBPSK Encoding

Bit Input	Phase Change
0	0
1	180

Table 5.3   DQPSK Encoding

Bit Input	Phase Change
00	0
01	90
11	180
10	270

("1" bit input). As indicated in Table 5.3, DQPSK operates on pairs of bits, encoding each pair into one of four predefined phase changes. Because DQPSK encodes two bits at a time, this technique is also referred to as dibit encoding.

Table 5.4 Physical Layer Throughput

Modulation Method	FEC Method	Physical Layer Throughput (Mbps)
DQPSK	¾ convolution code and Reed–Solomon code	13.78
DQPSK	½ convolution code and Reed–Solomon code	9.91
DBPSK	Convolution code and Reed–Solomon code	4.58
DBPSK	½ convolution code and Reed–Solomon code with each bit repeated four times	1.02

Forward Error Correction

To protect the integrity of data, the HomePlug specification defines four methods of forward error correction (FEC). Each method employs a different usage of convolution code and Reed–Solomon code or a different repetition of the bits in the codes, with different throughputs resulting from the use of each FEC method. Because FEC methods are associated with a modulation scheme, we need to look at both the modulation scheme and FEC method to determine the physical layer throughput. Table 5.4 indicates the physical layer throughput based on the modulation and FEC methods used.

Performance

From Table 5.4 it's obvious that at the physical layer throughput can vary dramatically. However, when 84 carriers are used per OFDM symbol with Ω convolution code and Reed–Solomon code, the physical layer provides a throughput of 13.78 Mbps (referred to as 14 Mbps in most literature). According to the HomePlug Powerline Alliance, this translates into a throughput of 8.2 Mbps at the MAC layer and a throughput of 6.2 Mbps at the TCP (Transmission Control Protocol) layer. Because Web pages are transported via TCP, let's perform a few computations to determine how long it will take to transmit a typical Web page over the electrical wiring in a home or office using HomePlug 1.0–compatible equipment.

The typical Web page consists of approximately 150,000 bytes of data. At 8 bits per byte, this results in the need to transport 150,000 bytes * 8 bits/byte, or 1,200,000 bits. At a throughput of 6.2 Mbps, the typical Web page can be transported in 1,200,000/6,200,000 = .19 seconds. Thus, the HomePlug specification is capable of delivering approximately five Web pages per second, which should be satisfactory for most home users and many small offices.

Adaptive Channel Allocation

To minimize the effect of noise on the electrical wires used for transmission, the HomePlug 1.0 specification defines an adaptive channel allocation scheme. Under this channel allocation scheme, each device samples the transmission on the electrical wiring and turns off heavily impaired carriers. This technique, also referred to as tone allocation, significantly reduces the bit error rate of the transmission. To further reduce errors, the HomePlug 1.0 specification defines the use of a concatenation of Viterbi and Reed–Solomon FEC, and sensitive frame control is encoded through the use of turbo product codes. The latter is a relatively new type of correction code that represents a significant improvement over previously developed error correcting codes. Turbo codes break down a complex decoding problem into a series of simple steps; each step is repeated until a solution is reached. According to several studies, turbo codes provide between 1.5 and 3.0 dB bit energy-to-noise improvement over standard Reed–Solomon and Reed–Solomon Viterbi error correction codes. Thus, the use of turbo codes to protect sensitive frame control provides a high degree of error-free re-creation of critical control information transmitted over electric wiring.

MAC Layer

Moving up the protocol stack, the HomePlug 1.0 specification defines the use of Carrier Sense Multiple Access with Collision Avoidance (CSMA/CA) for the MAC layer protocol. The use of CSMA/CA is supplemented by features that add priority classes, control latency, and enable support for QoS.

Encryption

Because electrical wiring represents a shared medium, HomePlug defines the use of the DES for privacy. Each station maintains a table

Table 5.5   Table of Encryption Keys and Encryption Key Select (EKS) Maintained by Nodes

EKS	Encryption Key	Meaning
0X00	0X08856DAF7CF58185	Default encryption key (unique for each node)
0X01	0X46D613E0F84A764C	Network encryption key (common for a logical network)
.	.	
.	.	
OXFF		

of encryption keys and associated encryption key select (EKS) values. The EKS values function as an index or identifier for each encryption key. Thus, a mechanism is required to inform each station of the EKS when it receives a packet. That is performed by including the EKS in the frame header, enabling a receiving node to select the applicable encryption key that will be used to decrypt the frame. Table 5.5 provides an example of the table of encryption keys and associated EKS values.

Under the HomePlug 1.0 specification, a series of nodes that encrypt transmission with a shared Network Encryption Key (NEK) form a logical network. To participate in a logical network, each node must have the NEK and associated EKS for the network. To simplify the setup of a logical network, most HomePlug devices are shipped with a default network password of 'HomePlug.' This setting generates a NEK using EKS 0X01, enabling out-of-the-box communications between nodes. In addition, most HomePlug 1.0 specification equipment used by this author enabled a change in the default network password to be distributed to all other devices in the logical network. Concerning those passwords, HomePlug equipment generates encryption keys automatically from the entry of ASCII passwords that a user programs into the device. Because the transmission protocol requires the movement of information within distinct frames, let's turn our attention to the frame formats supported by the HomePlug specification.

Frame Formats

As we previously noted, similar to other communications standards, the HomePlug 1.0 specification defines a MAC layer. At the MAC layer,

Long frame

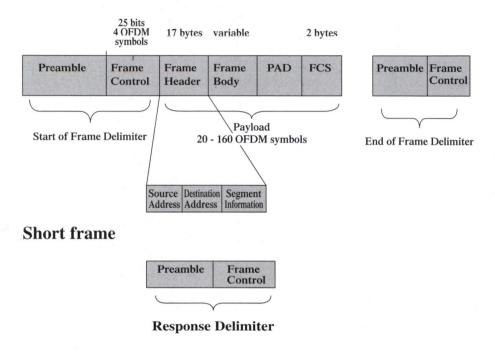

Short frame

Figure 5.2   Basic HomePlug frame formats.

HomePlug-compatible equipment uses a variant of the well-known CSMA/CA protocol. CSMA/CA defines how devices access the transmission medium. Because an understanding of the type of frames transmitted on a HomePlug network is vital to understanding how the CSMA/CA protocol works, we will examine the two basic frame formats defined by the specification.

Long and Short Frames

The top portion of Figure 5.2 illustrates the composition of the HomePlug long frame, whereas the lower portion of that figure indicates the fields in the HomePlug short frame. Because HomePlug uses a version of the CSMA/CA protocol, as you might expect, its frame formats resemble the formats of both Ethernet wired LANs and Ethernet wireless LANs. In fact, the HomePlug MAC layer was modeled to work with IEEE 802.3 Ethernet frames.

The HomePlug MAC layer uses a virtual carrier sense (VCS) mechanism and contention resolution to minimize collisions. The preamble field, which is common to both long- and short-frame formats, acts as a synchronizer for the receiver, which, upon receipt, attempts to recover the following frame control field.

Delimiters

The frame control field indicates if the delimiter, consisting of the preamble and frame control fields, is the start-of-frame, end-of-frame, or response delimiter. The start-of-frame delimiter specifies the duration of the payload that follows and is used in a long frame. The first 17 bytes of the payload of the long frame contains the frame header. This field consists of three subfields that contain the source address, destination address, and segmentation information. The source address indicates the sender and the destination address indicates the recipient. Segmentation information indicates if two or more frames are conveying linked information and were segmented due to packet length constraints. In comparison, the end-of-frame and response delimiters indicate the end of a frame and a response to a frame, respectively. For both of those delimiters, the frame control field indicates where the end of the transmission can be expected to occur. Thus, if a receiver is able to decode the frame control in a delimiter, it becomes capable of determining the duration of time that a channel will be occupied by the transmission. The receiver then sets its VCS based on the duration that the channel will be occupied. If the receiver cannot decode the frame control, it will then assume that a maximum-length packet is being transmitted and set the VCS accordingly. If it then receives an end-of-frame delimiter, the receiver may be able to correct its VCS.

Upon receipt of a unicast packet, the destination will acknowledge the packet by transmitting a response delimiter. If the source fails to receive an acknowledgment, it will assume that a collision caused the failure. In the event the destination has insufficient resources to process a packet, it will respond with a FAIL signal, whereas a NACK signal will be used to indicate that the packet contained errors that could not be corrected by the FEC code. As we will note, the ability to determine the duration that a channel is occupied minimizes the probability that a collision will occur.

Preamble

The preamble that prefixes each delimiter represents a form of spread spectrum signal that provides the receiver with knowledge that the start of a delimiter follows. The following frame control field, due to the importance of information it conveys, is encoded using a turbo product code. As previously mentioned, the use of a turbo product code represents an FEC mechanism that enables reliable data transmission to occur even at power levels several decibels below the level of noise.

Frame Control Fields

There are three types of frame control fields defined by the HomePlug specification, corresponding to the three frames shown in Figure 5.2. Each frame control field includes a type subfield and one to three additional subfields. Table 5.6 indicates the control subfields within each type of HomePlug frame. Because each HomePlug frame includes a delimiter, the frame control information fields listed in Table 5.6 are listed with respect to the type of delimiter that contains the control field.

Channel Access

Under the HomePlug specification, a combination of physical carrier sense (PCS) and VCS are used to determine if the medium is idle or busy and for how long. PCS occurs at the physical layer and indicates if the preamble signal is detected on the medium. In comparison, VCS is maintained at the MAC layer and is updated by information contained in delimiters that contain duration information. In addition, delimiters can contain information on which priority traffic can contend for the medium. Both PCS and VCS information is maintained by the MAC layer to enable the exact state of the medium to be determined.

Under the HomePlug channel access mechanism, the medium is first sensed to see if it is idle or busy. If the medium is busy, nodes will defer from transmitting until the medium becomes idle. When the medium becomes idle, nodes wait a randomly selected duration as part of the collision avoidance mechanism. If the node does not hear other traffic on the medium during the random duration, it can then transmit data. Similar to other CSMA/CA algorithms, the backoff algorithm used by the HomePlug specification selects a random integer between 0 and the contention window size. The algorithm functions as a mechanism to disperse the transmission times of frames that were queued as a result of needing to be retransmitted when the circuit was busy as well as provides a mechanism to ensure clients obtain access to the network in order of their priority. Because the priority mechanism

Table 5.6   Frame Control Information Fields

Type of Delimiter	Fields	Description
Start of frame (SOF)	Type	Can be an SOF with response expected or no response expected, depending on whether a short-frame delimiter is expected at the end of this long frame
	Contention control	When set to 1, prevents all nodes with packets of priority level equal to or less than the current long frame's priority from accessing the channel
	Frame length	Indicates the length of the payload in multiples of OFDM symbol blocks
	Tone map index	An index to the channel adaptation information stored at the receiver
End of frame (EOF)	Type	Can be an EOF with response expected or no response expected, depending on whether a short-frame delimiter is expected at the end of this long frame
	Contention control	Same meaning as for SOF delimiter
	Channel access priority	Indicates the priority of the current long frame
Response (Resp)	Type	Can be an ACK (positive acknowledgment), NACK (negative acknowledgment indicating faulty reception), or FAIL (negative acknowledgment indicating a lack of resources)
	Channel access priority	Indicates the priority of the long frame

used by the HomePlug specification provides a QoS capability as well as occurs within the contention window, let's probe deeper into the manner by which the HomePlug specification supports priority.

Table 5.7   HomePlug Priority Meanings

PRS0	PRS1	Meaning
1	1	Highest, priority 3, "voice," less than 10 ms delay
1	1	Priority 2, "video" or "audio" characterized by less than 100 ms delay
0	1	Priority 1, bulk data transfer and other background traffic
0	0	Lowest, priority 0, best effort traffic

Priority Support

The top portion of Figure 5.2 shows the transmission of a long frame. In response, the recipient will transmit a response delimiter consisting of a preamble and frame control field. The latter will indicate a good packet (ACK), errors detected (NACK), or a receiver busy condition (FAIL). Following the response delimiter is the contention resolution window, which contains two priority bit position indicators. Referred to as PRS0 and PRS1, they represent priority resolution intervals. When one node completes a transmission, the other nodes in the network that have queued packets to transmit signal their priority. In actuality, priority signaling occurs through the use of on/off keying, which enables the highest priority user to be determined if multiple users signal different priorities at the same time.

Through the use of two bit intervals, the HomePlug specification supports four priorities. Table 5.7 indicates the settings of PRS0 and PRS1 and their meanings.

As indicated in Table 5.7, the HomePlug specification provides four levels of priority. Those priority levels result from the mapping of the priority tagging method defined in the IEEE 802.11D specification, referred to as virtual LAN (VLAN) tagging. Under the IEEE 802.11D standard, seven user priorities are defined. Table 5.8 indicates the seven user priorities and the application class associated with each. In examining the entries in Table 5.8, we see that user priority A represents the highest priority, which can be characterized as a "must get there" situation necessary to support a network's infrastructure. User priority B represents voice that requires less than 10 ms of delay whereas user priority C represents video or audio that can tolerate up to 100 ms of delay. User priority D represents important applications that are subject to some form of admission control, and user priority

Table 5.8   IEEE 802.1D User Priority Defined Classes

User Priority	Application Class
A	Network control
B	Voice
C	Video or audio
D	Controlled load
E	Excellent effort
F	Best effort
G	Background

Table 5.9   Mappings of IEEE 802.1D Priorities to HomePlug Priority Classes

IEEE User Priority	HomePlug Priority Class
A	3
B	3
C	2
D	2
E	1
F	1
G	0

E represents the best-effort type of service that an information services organization would deliver to its most important customers. The sixth user priority, F, represents a best-effort application, such as LAN traffic, and the lowest priority, user priority G, represents bulk data transfers or other applications that can run on the network without impacting other applications.

Although the IEEE 802.1D standard defines seven priority classes, the HomePlug specification uses only four. The HomePlug specification mapped the seven IEEE 802.1D user priorities into four priority classes. Table 5.9 indicates the mapping and, when used in conjunction with Table 5.7, provides an indication of the evolution of the meanings associated with HomePlug priority classes.

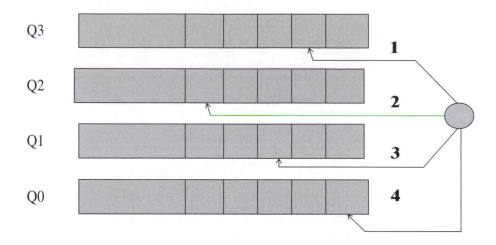

Rules:

If Q3 is nonempty, randomly select a packet waiting in the queue Q3

Otherwise, if Q2 is nonempty, randomly select a packet waiting in the queue Q2

Otherwise, if Q1 is nonempty, randomly select a packet waiting in the queue Q1

Otherwise, if Q0 is nonempty, randomly select a packet in queue Q0

Figure 5.3   HomePlug random packet extraction from queues.

HomePlug Queuing Model

From Table 5.9 we can note how the HomePlug specification maps the seven IEEE 802.1D user priorities into four priority classes. Each of the four priority classes that needs queuing will affect packet prioritization. The HomePlug specification defines a proprietary method that, although very complicated due to its many options, can be defined in terms of priority queues. Thus, we can simplify the HomePlug specification by noting that each priority class is provided with its own queue for network access. Once a packet is placed on the network it is allowed to flow to completion, with the resulting model becoming a nonpreemptive priority queue. However, instead of a "first-come, first-served" scheme to empty queues, the HomePlug specification results in packets selected randomly from each class. We can illustrate this extraction, so let's do so as well as indicate the rules associated with the packet extraction. The top portion of Figure 5.3 illustrates the packet formation within the four priority queues and the random extraction of packets from each queue. The lower portion of Figure 5.3 indicates the packet extraction rules.

Under the HomePlug specification, nodes that have the highest priority available in a network contend for access during the contention window period, whereas all other nodes defer an access attempt. All traffic with a priority less than the maximum availability traffic in effect wait until all higher priority traffic is transmitted.

Retries

Once a packet gains access to the network, there is a high probability it will flow to its intended recipient. However, if the packet is not received or if it is received in error, where one or more bits cannot be corrected by the FEC scheme, the packet will be retransmitted. Unfortunately, too many retries increase the delay of a packet, which can play havoc with the transmission of real-time information, such as digitized voice transported using VoIP. To avoid this, the original vendor who developed the equipment, Intellion, enabled users to adjust the number of times a packet can be retransmitted prior to it being dropped. Normally, the value is set to 6, but it can be changed on a node-by-node basis. By changing the retransmission value to 0 (no retransmissions) or 1, you can significantly improve the quality of real-time data flowing over electrical wiring. In fact, most implementations of VoIP simply discard erroneous packets because the delay associated with a retransmission can adversely affect the quality of reconstructed voice at a receiver. Under the HomePlug specification, a user cannot set a retry level, based on several vendor products I examined. However, because HomePlug supports up to four different priority classes based on the IEEE 802.1Q VLAN tag and includes a variable packet discard timer and maximum retry limit that is transparent to the user, this ensures that excessively delayed packets are discarded and reduces latency for higher priority traffic, such as voice and audio.

5.2 Working with Powerline Adapters

To provide readers with an indication of the ease of use of working with powerline adapters, I installed and used two Belkin USB powerline adapters. In this section we will examine the setup and communications of those adapters.

Overview

Belkin Corporation manufactures both Ethernet and USB powerline adapters, with the prefix referring to the manner by which the powerline adapter is cabled to a computer. Both types of powerline adapters are compatible with the HomePlug 1.0 specification, providing a maximum throughput of 14 Mbps at the physical layer, although the actual data rate at the MAC layer can be expected to be between approximately 40 and 50 percent of that obtainable at the physical layer.

Software Installation

The installation of software bundled on a CD-ROM with the hardware to include the operational setup of the device is relatively straightforward. Similar to most modern programs written for Windows operating system, the Belkin software uses the InstallShield Wizard to install the hardware on your computer. Figure 5.4 illustrates the initial welcome screen, which includes copyright information. Simply clicking on the button labeled "Next" calls up a new screen that will prompt you to enter your username and organization.

Figure 5.5 illustrates the customer information screen display when software is installed on a Windows XP–based computer. Because applications can be either restricted for the use of the person who installed the software or for all users of the computer, the installation program provides two options concerning the installation of the application. That is, you can choose to install the application for use by anyone who uses the computer, which is the default setting, or for exclusive use by the person who installed the application. This is shown in the lower portion of Figure 5.5.

Once you enter customer information and again click on a button labeled "Next," the program will display a screen indicating it is ready to install. This screen, which is shown in Figure 5.6, provides a summary of the settings that will be used to install the software. Because the software program appears to have been written to minimize user interaction, the default location of c:\Program Files\Belkin\ represents the location where the program will be installed. Unfortunately, there is no provision on the screen to change the default location or user information.

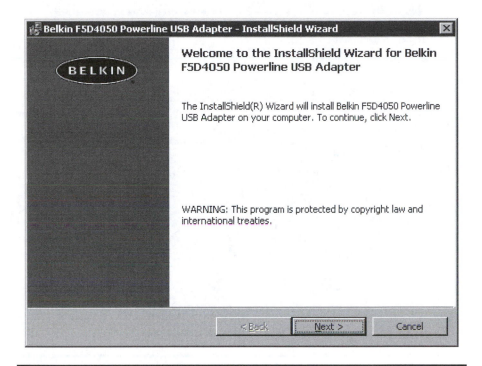

Figure 5.4   The initial Belkin welcome screen.

As the Wizard installs the program for the Belkin powerline USB adapter, a new screen will be displayed. This screen includes a status bar in the form of horizontal blocks that fill a rectangular area in tandem with the progress of the installation. Once the software is installed, another new screen display will appear that shows you how to complete the installation of the hardware. This screen, which is shown in Figure 5.7, both lists the three steps you need to perform as well as shows graphically what the steps refer to. If you carefully examine the person's left hand shown in Figure 5.7, you can note that the Belkin powerline USB adapter is smaller than a pack of cigarettes. Also note that the person's right hand holds a USB cable that is to be plugged into the adapter. Once this action occurs, Windows will automatically detect the adapter and launch the operating system's "Found New Hardware Wizard," which will associate the software and its hardware driver program with the hardware.

The Windows Found New Hardware Wizard will briefly display the message "Found New Hardware," indicating the hardware to be a USB–powerline bridge. Right after the display of the previously

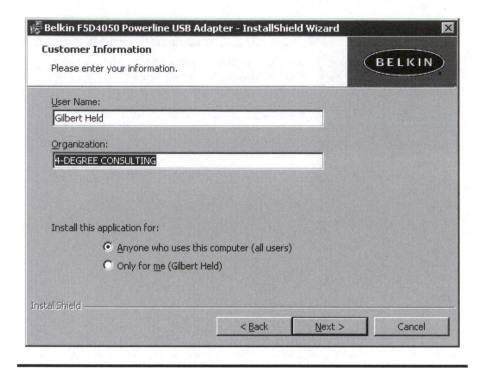

Figure 5.5   The Belkin customer information screen.

described message, you will be prompted to restart your computer to enable changes made to the Belkin powerline adapter to take effect. One interesting item to note concerning the installation of the Belkin software concerns the use of some spyware detection programs. The Belkin software adds a new layered service provider (LSP), which represents a method to intercept or modify network data. Although most Winsock LSPs are created by legitimate programs, on occasion they are manipulated by some software applications referred to as Winsock Hijackers. Thus, if your computer is running an antispyware program, it will more than likely detect the new LSP and provide you with several options to include ignoring its presence. Because the Belkin program is a legitimate application, you can safely select the ignore option on your antispyware display. Once you restart your computer, you can access the Belkin powerline configuration utility program in one of two ways — either through the Start menu by selecting Start>Programs>Belkin or by double-clicking on the Belkin icon displayed on your screen. Either action will result in the invocation of the Belkin powerline configuration utility program.

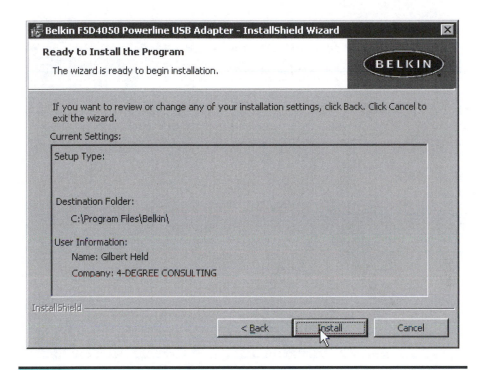

Figure 5.6 **The ready-to-install screen identifies the location where the program will be installed.**

Belkin Powerline Configuration Utility

The invocation of the Belkin powerline configuration utility program results in the display of a dialog box that contains five tabs. Figure 5.8 illustrates an example of the initial display of the configuration utility program with its default tab, labeled "Device," shown in the foreground.

Device Tab

In examining the Device tab shown in Figure 5.8, note that the "State" entry indicates the MAC address of the device to which your computer is directly cabled. In this example the MAC address is shown as 00:02:e3:38:4e:7e. As a refresher, the MAC address is a 48-bit address, which is subdivided. The first 24 bits or 6 hex digits represent a vendor identifier and are assigned by the IEEE. The last 24 bits or 6 hex digits

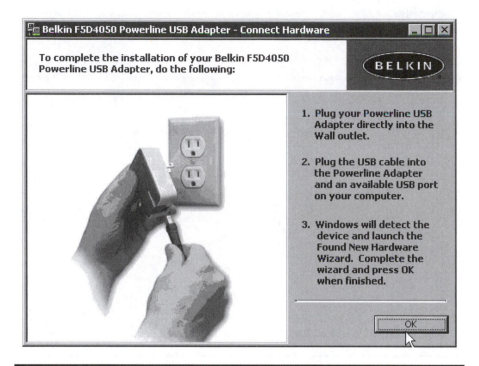

Figure 5.7 The Belkin "Connect Hardware" screen shows users how to connect the USB cable to the powerline USB adapter.

are assigned by the vendor and ensure that each adapter has a unique MAC address. The window located in the middle of the dialog box includes a heading labeled "Device." This will initially show the same powerline device as that shown in the rectangular box labeled "State." However, if there are multiple powerline devices connected to your computer, you will be able to select the device you would like to configure by highlighting it in the list and clicking the button labeled "Connect." In addition, you can use the "Refresh" button to scan for powerline adapters directly connected to your computer. At the bottom of the dialog box you will note a rectangular area labeled "Link Quality." This bar indicates the signal strength at the time of the scan. When you first install one adapter, the link quality will be displayed as "Poor" because the single adapter cannot communicate with another adapter to determine the quality of the connection. As we will note, when we set up a second powerline adapter, which has the first to communicate with, the link quality will appear as excellent.

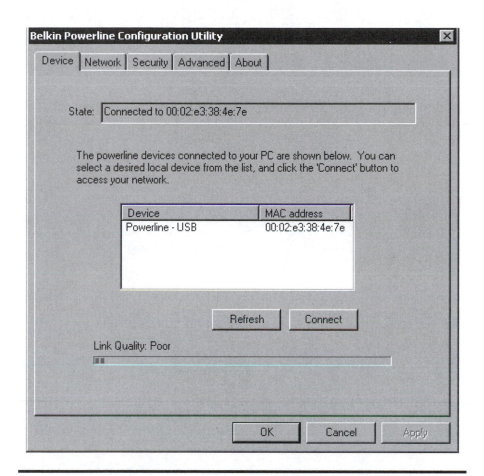

Figure 5.8 **The Belkin powerline configuration utility program.**

Network Tab

When you select the Network tab, you are able to view the MAC address and data rates of all remote powerline devices that make up your network. Figure 5.9 illustrates the Belkin powerline configuration utility program with the Network tab positioned in the foreground of the dialog box. In this example one remote powerline device was located on the network and the data rate is indicated as 6.35 Mbps. Because other powerline-compatible devices could be added to the network after the Network tab is selected, the lower portion of the display includes a "Scan Powerline Network" button that, when

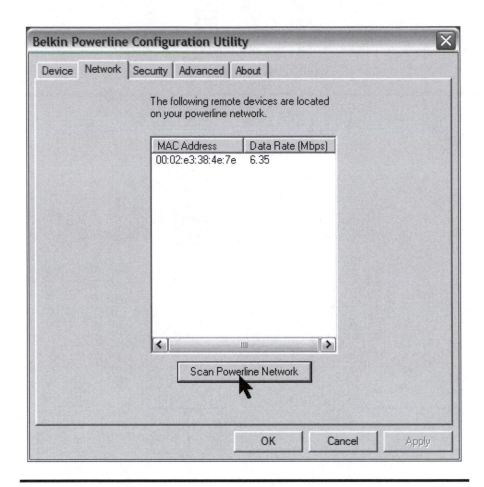

Figure 5.9   The Belkin powerline configuration utility Network tab.

selected, results in the scanning of the network for all devices that may be operational at the time of the scan.

Security Tab

As mentioned earlier in this chapter, most HomePlug-compatible devices are configured with a default network password of "Home-Plug." The Belkin powerline adapters are no exception to the use of this default setting, as we will note.

Figure 5.10 illustrates this in the Belkin powerline configuration utility program, with its Security tab positioned in the foreground of the display. This tab provides you with the ability to change the default network password as long as you set up each device on the network

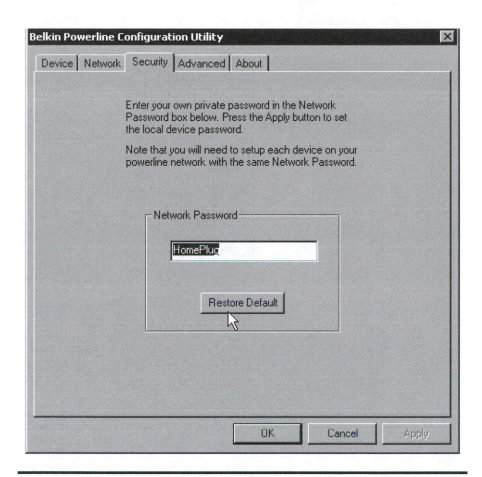

Figure 5.10   The Belkin powerline configuration utility Security tab.

with the same new password. Also note in examining Figure 5.10 that after you change the network password you can restore it to its original default value by simply clicking on the button labeled "Restore Default."

Advanced Tab

Because it can be time consuming and perhaps even physically taxing to move to various locations within a home or small office to reprogram the network password and distributed devices, the Belkin powerline configuration utility provides a mechanism to set the password on remote devices. To do so, you would click on the Advanced tab in the powerline configuration program, resulting in a display similar to that shown in Figure 5.11.

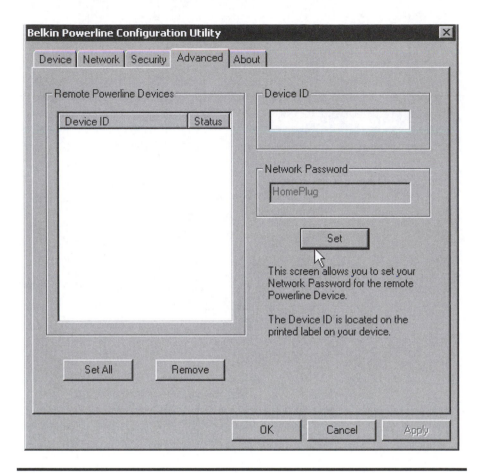

Figure 5.11   The Belkin powerline configuration utility Advanced tab.

In examining the Advanced tab screen shown in Figure 5.11, note that setting a network password for a remote powerline device requires the user to first enter the device ID into an applicable box on the right-hand side labeled "Device ID." The device ID is located on a printed label on each powerline device and represents a 16-digit alphanumeric value consisting of four groups of numbers separated from one another by dashes. The device ID should not be confused with the MAC address of the device, with the latter representing a 12-digit hexadecimal number.

Returning our attention to Figure 5.11, once you enter a remote device ID and press the "Set" button, the current network password will be applied to the remote device. The device ID will then be added to the "Remote Powerline Devices" text list box, with its status indicated in the box. Below the text box you will note the buttons labeled "Set

All" and "Remove." You would click the "Set All" button to apply your current network password to all devices listed in the text box or click the "Remove" button to remove a previously highlighted device ID from the text box.

5.3 HomePlug AV

The HomePlug Powerline Alliance several years ago recognized the need for supporting audio and video multistream entertainment via in-building ac electrical wiring. The result of this recognition of the need to support the transmission of evolving multimedia applications, to include high-definition television (HDTV), was the beginning of a new specification referred to as HomePlug AV.

Overview

The HomePlug Power Alliance issued its market requirements document (MRD) for the HomePlug AV specification during February 2003. The objective of the evolving specification was to support high-quality video distribution, secure connectivity, and robust control of QoS, and be compatible with existing HomePlug 1.0 devices. The MRD was developed as a guide for vendors submitting technical proposals to the HomePlug Technical Working Group, which is the organization tasked with selecting HomePlug AV's base technology. As of late 2005, the completion of the AV specification was expected to occur in early 2006; however, certain aspects of the emerging specification were known.

Operational Characteristics

In addition to being compatible with the HomePlug 1.0 specification, the emerging HomePlug AV specification adds advanced functionality to move digital audio and video data streams over electrical wiring. In doing so, the HomePlug AV specification is designed to provide a 200-Mbps data rate, which is sufficient to transport multiple HDTV programs around a home. Other functions included in the preliminary HomePlug AV specifications are an advanced physical layer that provides near-capacity throughput performance as well as robust communications over noisy power lines, a highly efficient MAC layer that includes both scheduled access via time division multiple access (TDMA) with QoS guarantees and contention access via the use of

CSMA/CA, and advanced network management capabilities. The latter facilitates plug-and-play operations, which will enable both users and service providers to easily make Home Plug AV equipment operational.

Compatibility

One of the more interesting aspects of the HomePlug troika of existing and evolving specifications is the fact that they are planned to inter-operate with one another. To accomplish interoperability, the Home-Plug AV evolving specification includes co-existence modes, which provide backward compatibility with HomePlug 1.0 devices and for-ward compatibility with devices that will eventually operate according to the BPL specification.

5.4 IEEE Wireless LANs

In concluding this chapter we will turn our attention to the series of IEEE wireless LAN standards developed by the 802.3 committee of that organization. The reason for concluding this chapter by focusing our attention on the IEEE 802.11 series of wireless LAN standards results from several BPL developers using wireless LANs as an alternative to powerline modems for communications into homes and small offices. Because one access point can serve several homes and offices, the economics associated with providing an over-the-air transmission facil-ity for the last few hundred feet may be more attractive than using several BPL modems.

Overview

The original IEEE 802.11 standard for wireless LANs was finalized during 1997. This standard defined three different physical layer spec-ifications. Two specifications are radio frequency (RF) based, operating in the 2.4-GHz frequency band, and the third is for infrared commu-nications. The two RF-based specifications defined the use of direct sequence spread spectrum (DSSS) and frequency hopping spread spectrum (FHSS) at data rates of 1 and 2 Mbps. Similarly, infrared communications were defined for data rates of 1 and 2 Mbps. Although several vendors have manufactured 802.11 RF-compatible products, to this author's best knowledge the infrared version of the IEEE 802.11 wireless LAN has not been marketed.

Both DSSS and FHSS represent transmission techniques originally developed for military applications. Referred to as "wideband" transmission, both transmission methods made jamming difficult. FHSS results in each portion of a transmission hopping from frequency to frequency, whereas DSSS results in data bits being spread over a range of frequencies. It wasn't until the 1980s that their ability to withstand jamming was also viewed as a mechanism to boost throughput as well as ensure that a signal reached its destination.

The Wireless LAN Infrastructure

The IEEE 802.11 specification defined a series of wireless LAN components and how they interoperated with one another. The basic service set (BSS) represents the smallest building block of an 802.11 wireless LAN. The BSS is formed by two or more wireless LAN stations executing the same MAC protocol and using the same shared media. The connection of two BSSs occurs via a distribution system (DS), which, although the transmission method for the DSS is not defined, can be a wired backbone LAN, another wireless LAN, or based on a yet-to-be-developed technology. The connection of two or more BSSs via a DS forms an extended service set (ESS). Figure 512 illustrates the relationship between BSSs, a DS, and an ESS.

Communications Methods

There are two methods by which stations can communicate within an IEEE 802.11 wireless LAN. The first method enables stations to communicate directly with one another and is referred to as ad hoc transmission. The second method requires all communications to flow through an access point and is referred to as infrastructure communications. Because the access point can be connected to a cable or DSL (Digital Subscriber Line) modem or another network, most wireless LANs use the second method of communications.

Medium Access

Wireless LAN stations use the CSMA/CA protocol to gain access to the medium. Two different MAC techniques are defined —PCS and VCS. Under the PCS method, a station that has data to transmit first listens to the medium to make sure that the transmitting station has completed its activity. Then, the station with data to transmit waits a short period

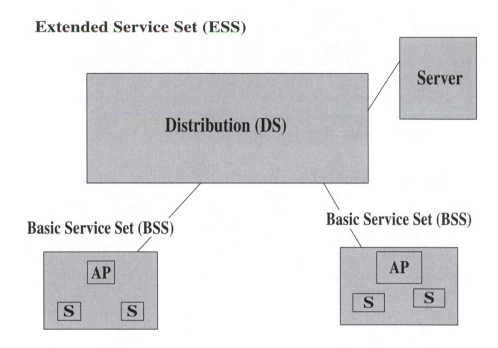

Legend: AP - access point, S - station

Figure 5.12 The wireless LAN infrastructure.

of time defined by a field in a previously transmitted frame. Each frame is transmitted using a stop-and-wait Automatic Repeat-reQuest (ARQ), so by waiting, the listening station can detect that the transmitting station is finished and can then begin its transmission. Both ACK and NAK signals responding to a prior frame are transmitted a short time after a frame is received, whereas stations wishing to transmit a frame wait a little longer to ensure that no collisions can occur.

The second method of medium access is the VCS method. When a station is near the edge of the transmission range of an access point and another station is at the opposite edge of the transmission range of the access point, both stations can transmit to the access point, but more likely than not, they cannot detect each other's signal. This condition is referred to as a "hidden node" problem and results in the inability to use the PCS method of medium access. The VCS method solves the hidden node problem by having a transmitting station first transmit a request-to-send (RTS) frame to the access point. If the access point responds with a clear-to-send (CTS) signal, the station with data to send can then begin its transmission. By default, most wireless LAN hardware is set up so that the use of RTS and CTS signaling is disabled.

IEEE 802.11 Extensions

The original IEEE 802.11 series of wireless LAN specifications did not achieve a significant degree of utilization. Perhaps the main problem was their maximum operating rate of 2 Mbps. Recognizing this limitation, the IEEE developed a series of extensions to the original 802.11 specification. These extensions altered the frequency band, the data rate, and the modulation method from the original specification. The four extensions are referred to as 802.11a, 802.11b, 802.11g, and 802.11n.

802.11a

The 802.11a extension defines the use of the 5-GHz frequency band and uses OFDM to obtain a maximum data rate of 54 Mbps. Because high frequencies attenuate more rapidly than low frequencies, the transmission range of stations compliant with the 802.11a extension is narrower than stations operating in the 2.4-GHz frequency band specified for RF operations by the original specification. Due to its relatively short range in comparison to other IEEE extensions, to this author's knowledge, no BPL field trials use this technology.

802.11b

A second extension to the 802.11 specification is referred to as the 802.11b extension. The 802.11b extension specifies the use of DSSS at data rates of 11, 5.5, 2, and 1 Mbps. Because the 802.11b extension operates in the 2.4-GHz frequency band, its range is between two and four times that of 802.11a devices; high frequencies attenuate more rapidly than lower frequencies. Although some vendors market dual 802.11a/b devices, such devices operate as either an 802.11a or an 802.11b device at a particular point in time. Examination of five BPL field trials involving the use of wireless LANs for transmission into homes revealed that each trial used 802.11b equipment.

802.11g

The 802.11g extension can be viewed as an extension to the 802.11b extension. The 802.11g extension also supports operations in the 2.4-GHz frequency band and is backward compatible with the 802.11b extension. In addition, it supports the use of OFDM in the 2.4-GHz frequency band, which enables a maximum data transmission rate of 54 Mbps. Since its introduction during 2003, several vendors have developed 802.11a/b/g products, which enable a station to become compatible with three different types of access points.

Currently, the 802.11g extension provides for the support of more simultaneous users and a greater signal range than the older extensions. This will more than likely change when the 802.11n specification is finalized during 2006. Although I was not able to locate the use of 802.11g technology in BPL field trials, I expect vendors to migrate to this technology because of the higher data rate and extended range of the technology in comparison to the 802.11b specification.

802.11n

The 802.11n specification builds on previous 802.11 standards by adding a multiple-input, multiple-output (MIMO) capability.

During 2005, several vendors introduced MIMO products that can be considered proprietary because the specification will more than likely be completed during the following year. MIMO is based on the use of multiple antennas and exploits such phenomena as multipath propagation to increase throughput. True MIMO requires a 2×3 antenna configuration, with two antennas used for transmission and three used for reception. This explains why several vendors introduced "pre-n" products with three antennas during 2005. Such products are designed to work according to the draft IEEE 802.11n specification and provide a throughput of up to 108 Mbps at distances up to 300 ft. However, because the final specification may differ from the draft specification, the possibility exists that current pre-n products may lack an inter-operability capability with products built to comply with the finalized specification. In spite of this problem, I expect the IEEE 802.11n specification to eventually become the most popular wireless LAN due to its higher throughput and extended transmission range. As more standardized 802.11n products reach the market, it is highly likely that BPL providers who use wireless LAN technology in place of the low-voltage line into homes and small offices will begin to use products compliant with the 802.11n specification once that specification is finalized during 2006.

Chapter 6

Equipment Vendors and Field Trials

In previous chapters in this book, we examined power line operations from generators to the home or office, how data transmission can occur over power lines, and the interference caused by such transmission. Although we also examined broadband over power line (BPL) equipment when discussing the major architectures used by vendors, for the most part our reference to hardware was for generic devices. In this chapter we will probe much deeper into the operation of BPL hardware by turning our attention to equipment vendor product offerings. Thus, in the first section of this chapter we will focus our attention on the operational capability of BPL equipment vendors. Although we can reasonably expect additional vendors to manufacture BPL equipment as usage expands, this section should provide readers with a solid background concerning the product offerings of several vendors at the time this book was written. However, readers should understand that the products mentioned are representative of current offerings and the evolution of technology can be expected to result in some changes to product operations. In the second section of this chapter, we will turn our attention to several BPL field trials, examining the structure or architecture used to deliver a data transmission capability as well as how vendor equipment was used. All product offerings and field trials discussed in this chapter will be presented in alphabetical order.

Concerning the second section of this chapter, because one field trial resulted in the first deployment of BPL technology in the United States, we will examine both the field trial and the deployment associated with the City of Manassas, Virginia, as an entity.

6.1 Equipment Vendors

As previously mentioned, our examination of equipment vendors is presented in alphabetical order. Because in many instances BPL equipment provides transmission into the home or office over low-voltage lines, it becomes necessary to use power line modems at the subscriber location. Thus, in this section we will also examine a few vendor products developed for in-home use that enables communications not only through the home or office but, in addition, over low-voltage lines to BPL equipment installed by the electric utility on their infrastructures. As we investigate equipment vendors, we will note that although there are some similarities between several product lines, different terms are used to reference similar-functioning equipment.

Ambient Corporation

At the time this book was prepared, Ambient Corporation was the only publicly traded company located in the United States dedicated solely to manufacturing power line communications (PLC) technology products. The company's PLC technology was developed to allow utilities to deploy a communications network over their existing power line infrastructure of low- and medium-voltage lines. To accomplish this task, Ambient defined a two-part system architecture. The first part overlays a utility distribution network, whereas the second portion of the architecture consists of the backhaul data network, which connects the PLC network to the Internet and the utility data center. The latter represents a location where the utility can manage the network as well as enable and disable the operation of equipment used by subscribers.

The architectural design developed by Ambient enables the sequential installation of devices to correspond to customer demand. Because this architecture minimizes installation cost, it also shortens the time for a utility to recover its cost through monthly subscriber revenues.

The Ambient network architecture defines the use of network and physical layer components. Network layer components are connected to the utility distribution network to route data traffic over power lines, whereas physical components are attached to utility power lines to transmit and receive data over those lines as well as to address utility devices, such as transformers, that pose a challenge to the flow of high-frequency data signals.

Network Layer Components

Under the Ambient network architecture, electronic devices referred to as "nodes" are deployed at selected locations in the utility distribution network. This deployment in effect results in an overlay of a communications network onto the utility power lines. Such devices, or nodes, are located at substations, transformers, and customer premises, and at selected distances on power lines. We will examine four distinct network layer components defined by Ambient.

S-Node

The S-node is located at a utility substation and can be considered to represent a substation node. Because a substation has many feeder power lines routed into neighborhoods, it represents an optimum location from which the data flow over feeder lines can be connected to a backhaul network. Because some utilities may have fiber connections to substations whereas other utilities may lack such connections, the Ambient S-node was designed to support both fiber and metallic connections, including T1 connections. In addition, the S-node also supports multiple power line interfaces, which enables the device to connect to all of the feeder lines at a substation. This is accomplished by the ability of the S-node to support multiple power line modems.

In addition to supporting multiple feeder power lines, the S-node also supports multiple connections to backhaul facilities. Thus, the S-node can be used to connect to a backhaul via a fiber or metallic line as well as to a utility data center or to other networks.

X-Node

A second electronic device defined under the Ambient network architecture is the X-node. The X-node is located at a transformer and is also referred to as a transformer node. This node is designed to transfer data between medium- and low-voltage lines as well as to function as a repeater along the medium-voltage line.

R-Node

A third Ambient network layer component is the R-node. This device, which is also known as a repeater node, can be considered to represent a modified X-node that lacks a connection to a low-voltage line. Thus, the R-node is designed for use on medium-voltage lines to alleviate the effect of high attenuation of data transmission signals.

GW-Node

A fourth Ambient network layer component is the GW-node, which is also referred to as a gateway node. The GW-node is installed at the customer's premises. This device contains a PLC interface to the low-voltage line and Ethernet and plain old telephone service (POTS) interfaces to support an in-home data network as well as Voice-over-IP (VoIP) telephony. The GW-node is designed as a modular device that can enable support for additional utility services to include meter reading as well as in-home wireless networking. Thus, the GW-node can be viewed as the equivalent of a Swiss Army knife due to its support for a variety of network interfaces.

Physical Layer Components

Under the Ambient network architecture, the physical layer represents devices that are connected to power lines to add or extract data signals. There are three physical layer interface components: modems, coupling devices, and line conditioners.

Modems

PLC modems are similar to conventional modems in that they modulate and demodulate data. Modems used by Ambient are positioned at locations within the utility power distribution network where they can effectively receive signals from other modems. Due to noise and other impairments, modems include an error correction capability.

Couplers

Couplers are devices that enable the transmission and reception of modem signals from the power line. Ambient uses a proprietary passive inductive coupler that facilitates coupling data signals to and from power lines. Ambient was issued a patent during late 2002 for the technology used in its coupler.

Ambient inductive couplers are designed to operate in both overhead and underground systems. The medium-voltage overhead coupler provides signal transmission and reception on power lines with 2.4 kV to 35 kV. This coupler is installed directly onto an overhead medium-voltage line at a substation; it connects with an S-node near an electric pole on which one or more distribution transformers are mounted. In comparison, the medium-voltage underground coupler is used to provide an attachment to the phase wire on the primary side of pad-mounted transformers. Similar to the overhead coupler, the underground coupler is capable of operating with voltages up to 35 kV.

Line-Conditioning Devices

The third physical layer component is the line-conditioning device. This device operates on both medium- and low-voltage distribution lines and functions to either pass or block signals based on the composition of predefined signals. Thus, the line-conditioning device functions as a mechanism to subdivide or sectionalize the utility distribution network. Line-conditioning devices are typically installed near such utility devices as switches and capacitor banks.

Amperion, Inc.

This Lowell, Massachusetts, based company specializes in medium-voltage PLC technology. Amperion was founded in 2001 with initial funding of $12.5 million and represents a joint venture backed by Cisco Systems, Redleaf Group, and American Electric Power; the latter a major electric utility. The company develops network hardware and software to enable the delivery of broadband communications over the electric utility infrastructure.

Product Lines

At the time this book was researched, Amperion offered three product lines. The firm's Lynx enables PLC service on an underground utility infrastructure, and Amperion's Griffin and Falcon product lines enable service on an overhead utility infrastructure. Each product line consists of an injector, an extractor, and a combined repeater–extractor. The three product lines are marketed under the Amperion Connect Solution moniker, providing up to 24 Mbps throughput per injection point.

The Injector

The injector represents the interface between the network access connection and the medium-voltage feeder line. Usually, one injector is installed per medium-voltage feeder. At the network access connection, the injector supports fiber and metallic, with the latter including T1 and T3 connections.

The injector converts the signal received from the network access connection to a signal suitable for transmission on the medium-voltage wire. Once on the wire, the signal will either be repeated to extend its transmission or be extracted to deliver bandwidth to the subscriber.

The Extractor

The extractor is the device that connects the communications network overlaid on the medium-voltage lines to the subscriber. Extractors can be used to deliver bandwidth to residences, businesses, wireless towers, and utility devices.

The Repeater–Extractor

The Amperion repeater–extractor provides both a repeater and an extractor capability. The repeater portion of the device first receives a signal transmitted by an injector or another repeater and performs an error correction operation to minimize the effect of noise prior to amplifying the signal. The device also functions as a router, decoding received IP datagrams and routing data based on the destination address in the datagram.

Amperion's Griffin product is mounted on a utility pole, whereas the Falcon sits directly on the medium-voltage line. The Griffin is powered through a 110-volt secondary line connection. In comparison, the Falcon products can be powered inductively from the medium-voltage power line or from a pole-mounted transformer.

Figure 6.1 illustrates the connection of the Griffin extractor to the medium-voltage power line. Both the Griffin and the Falcon support the IEEE 802.11b wireless interface, enabling one extractor to serve multiple users via wireless LAN communications. The insulated coupler represents the signal interface between the unit and the medium-voltage feeder. At the top of the enclosure shown in Figure 6.1 you

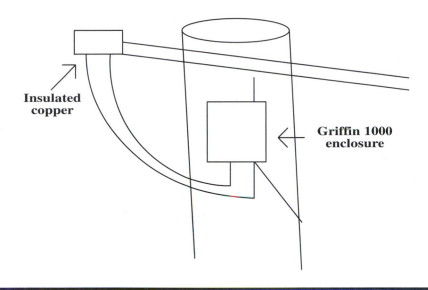

Figure 6.1   The Amperion Griffin 1000 provides data rates up to 24 Mbps per medium-voltage feeders.

will note the wireless antenna, which enables subscribers within over-the-air range of the Griffin 1000 to obtain broadband communication without having a power line modem in the home or office.

Corinex Communications Corporation

This Vancouver, British Columbia, headquartered company claims to be the world's leader in deploying networking solutions based on power line, cable, phone line, wireless, and VoIP technologies. Corinex Communications is a privately held company that was founded in 1989. It performs its R&D engineering in Bratislava in the Slovak Republic, and manufacturing occurs in China and Taiwan. The company currently markets several power line products for use in the home and office. Those products include a Powerline Ethernet Wall Mount, a Powerline Ethernet Adapter, and a Wireless to Powerline Router that uses OFDM technology and complies with the HomePlug 1.0 standard that supports data rates up to 14 Mbps.

Powerline Ethernet Wall Mount

The Powerline Ethernet Wall Mount has the shape of a miniature mouse, with dual plugs for insertion into an electrical outlet. This device uses the 110/120- or 220/240-V electrical outlet as a medium for communications at a data rate up to 14 Mbps. The Powerline Ethernet Wall Mount includes an Ethernet jack that enables the device to be cabled to an Ethernet slot or card in a PC and is bundled with a CD. The CD contains a setup tool that enables the device to be programmed with private and personal encryption keys to prevent unauthorized persons from intercepting and understanding data flowing over the electric wires in a home or office. All devices in a network must be programmed with the same network encryption key, including the Corinex Wireless to Powerline Access Point and other Corinex power line products.

Each Powerline Ethernet Wall Mount has a transmission range of approximately 200 meters. Various versions of the device come equipped with U.S., U.K., and Euro ac plugs as well as models for 100/120- and 220/240-V ac operations.

AV 200 Powerline Ethernet Adapter

The Corinex AV 200 Powerline Ethernet Adapter can be considered to represent an enhanced version of the vendor's Powerline Ethernet Wall Mount. The AV 200 includes a 10/100 BASE-T fast Ethernet interface and is capable of transmitting data at distances up to 300 meters at data rates up to 200 Mbps. Because the AV 200 supports the IEEE 802.1P and 802.1Q standards, it can support priorities and VLAN (virtual local area network) operations, including the ability to stream several high-quality video signals, such as high-definition television (HDTV) while providing high-speed Internet access.

Wireless to Powerline Router G

The Corinex Wireless to Powerline Router G provides home and business users with the ability to connect computers by utilizing both electrical wiring and wireless over-the-air transmission. This router supports the IEEE 802.11g standard, providing a data rate up to 54 Mbps for over-the-air transmission. In addition, the router also has an Ethernet WAN (wide area network) port for connection to a DSL (Digital Subscriber Line) or cable modem, an Ethernet LAN port for connection to stationary equipment, and a HomePlug 1.0 powerline interface for connection to other devices via electrical wiring at data rates up to 14 Mbps at distances up to 200 meters.

The Corinex Wireless to Powerline Router G can be used to form the backbone of a home or small office network. Because it supports wireless and wired LAN connections as well as access via electrical wiring within the home or small office, it can function as a central point of network connectivity. Thus, transmission limitations associated with wireless LANs used in buildings can be overcome by using power lines to enable out-of-reach devices to communicate via electrical wiring. In addition, this device includes support for legacy WEP (Wired Equivalent Privacy) as well as the more modern IEEE 802.11i security standard. Because the router functions as both a router and an access point, it includes support for network address translation (NAT), Dynamic Host Configuration Protocol (DHCP), Routing Information Protocol (RIP) versions 1 and 2, as well as static routing. For management, the Wireless to Powerline Router G includes a SNMP (Simple Network Management Protocol) agent.

Wireless to Powerline Access Point

The Corinex Wireless to Powerline Access Point can be viewed as a similar but less functional product than the Wireless to Powerline Router G. This is because the Wireless to Powerline Access Point does not include a routing capability and its wireless interface supports the IEEE 802.11b standard, which is limited to a maximum data transfer rate of 11 Mbps. However, because this device supports both powerline and wireless communications, it enables communications with other computers both over the air and via electrical wiring. The Powerline Access Point is compatible with the HomePlug Powerline Alliance Industry Specification 1.0.1 in that it supports transmission over electrical wiring at distances up to 200 meters at data rates up to 14 Mbps.

Other Powerline Adapters

To round out its product line, Corinex produces a family of powerline adapters that can be used to connect to computers with different interfaces. Those products include a Powerline Ethernet Adapter and a Powerline USB Adapter. As their names imply, the Powerline Ethernet Adapter permits a computer with an Ethernet point to become a participant on a power line network and the Powerline USB Adapter provides similar functionality for computers with an USB connector.

Current Communications Group

Current Communications Group of Germantown, Maryland, has three subsidiaries involved with partnering with utilities to provision technology and equipment that enable broadband transmission over power lines. The subsidiaries of Current Communications Group include Current Technologies, which develops technology and equipment; Current Link, which establishes joint ventures with utilities to offer BPL capability and services; and Current International, which is tasked with seeking international opportunities.

As an equipment developer, Current Technologies provides five principal components that are employed to create a BPL overlay on an existing electric utility infrastructure. Those components include the CT Coupler, CT Backhaul-Point, CT Bridge, Powerline modem, and network management system.

CT Coupler

The CT Coupler functions as a transformer bypass, allowing broadband signals to travel to the CT Bridge and CT Backhaul-Point. Current Technologies has a stated goal of 20 minutes for the installation of a CT Coupler.

CT Backhaul-Point

The CT Backhaul-Point functions similar to a multiplexer, aggregating traffic from multiple CT Bridges. The aggregated traffic can then be transmitted via a powered Ethernet or fiber-optic port interface onto the backhaul network, which connects to an Internet Service Provider (ISP). The CT Backhaul-Point uses a CT Coupler to communicate across medium-voltage lines.

CT Bridge

The CT Bridge is installed next to a transformer and functions as a gateway between low- and medium-voltage lines. The CT Bridge provides a symmetric transmission capability to a group of homes served by a transformer. The CT Bridge provides routing, security, subscriber management, and a DHCP capability. The device also includes a powered Ethernet option, which allows the device to be interfaced to an access point and functions as a WiFi hot spot.

Powerline Modem

The Powerline modem, although a critical component of the Current Communications BPL system, represents a third-party product. The Powerline modem can be obtained from any vendor that manufactures a HomePlug 1.0 standard modem, such as Belkins, Linksys, Netgear, and Siemens. Under the HomePlug 1.0 standard, Powerline modems have a raw data rate of 14 Mbps, with actual throughput between 2 and 6 Mbps.

Depending on the manufacturer, customers can select a Powerline modem that supports common interfaces. Currently available interfaces include a USB port, a 10/100-Mbps Ethernet port, and a wireless 802.11 interface.

CT View

The last component of the Current Communications network is its CT View network management system. Referred to as CT View, this network management system provides configuration, fault, accounting, performance, and security-related functions. CT View includes an auto-discovery feature that enables this network management system to locate each device once it is installed. In addition, CT View supports alarm generation to alert operators to abnormal conditions and constructs a database that enables operators to examine various device parameters.

Figure 6.2 illustrates the relationship between Current Communications' five principal hardware components. In examining Figure 6.2, note that Current supports BPL directly over low-voltage lines into homes or via IEEE 802.11b wireless LAN communications. The former requires the home user to install a Powerline modem whereas the latter allows the use of a wireless adapter connected to a user's PC.

DS2

DS2 is a Valencia, Spain, supplier of semiconductor chipsets and software for power line communications. The company markets a series of chipsets that supports 200-Mbps data transmission over power lines. The firm's DSS9010 represents a chip optimized to support high-speed video and data communications over existing "domestic power cables," a term used by DS2 to reference in-building electrical power lines. The DSS9010 chipset supports the use of OFDM modulation to

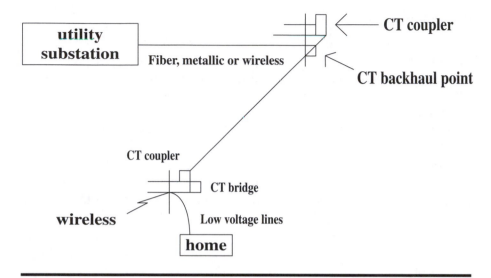

Figure 6.2   Relationship of Current Communications' hardware components.

provide a 200-Mbps data transmission capability and provides the foundation through its support of quality of service (QoS) for incorporation into modems as well as for use in other types of infrastructure equipment.

In addition to supporting QoS, the DSS9010 supports 3DES (Data Encryption Standard) encryption, multicast transmission, and remote management with SNMP. The DSS9010 also contains an interface for 100 BASE-T Ethernet, which enables the chipset to be used with a variety of devices that can be connected via the common and near-ubiquitous Ethernet plug.

Enikia

Located in North Plainfield, New Jersey, Enikia is a privately held company that develops technology to provide Ethernet-speed communications in homes over the existing power line network. Enikia, whose name derives from words in Greek meaning "at home," is a founding member of the HomePlug Powerline Alliance.

aiPACK

The firm's first product was developed for PLC testing. That product was called aiPACK, short for Access and Inhome Powerline Characterization and PLC Loop Testing Toolkit. aiPACK provides testing and characterization for utility companies with preliminary PLC implementations designed for operating on the low-voltage segment of the power line network, enabling measurements at the subscriber's premises.

Power Bridge

A recently introduced Enikia product is its Power Bridge. The Power Bridge uses an Enikia-developed semiconductor chip to provide both low-latency and QoS capability. The Power Bridge is plugged into a home or office 110-V ac power line to both derive power for the unit as well as use coupling circuits for transmission and reception of data. The Power Bridge includes two RJ45 10 BASE-T Ethernet ports as well as two optional POTS connectors, the latter supporting the direct connection of two conventional telephones. The front of the Power Bridge includes two 16-character LCD display lines, enabling monitoring and the display of status messages.

Through the use of the Enikia Power Bridge, customers of power line providers can obtain the ability to support such real-time services as telephony via VoIP, audio, and video. With a transmission distance of up to 300 ft, this device can be used at the subscriber premises to provide support for real-time applications through the low-voltage power line connection of the utility providing electric service.

Main.net Communications Group

Main.net is an Israel-based company that has developed power line communication solutions that combine both home networking and the aspects of PLC into a unified communications solution used by several electric utilities. Marketed under the name PLUS, as an acronym for Power Line Ultimate System, the PLUS system enables high-speed Internet, telephony, and home networking as well as supporting such additional services as burglar and fire alarms, surveillance cameras, energy management, and asset tracking. The Main.net group includes Main.net Communications Ltd., located in Israel; Power Plus Communications AG, which represents a joint venture with the German power utilities MW and ABB; and its American subsidiary, Main.net Power Line Communications, Inc., which is located in Virginia. Main.net Power Line Communications is the first equipment vendor to have its products move from a field trial to deployment in the United States. When we discuss field trials in the second section of this chapter, we will note how this vendor's products are used by the City of Manassas, Virginia, to provide a citywide BPL capability.

Main.net PLUS Solution

Main.net's PLUS solution was developed to facilitate the installation of equipment necessary to overlay an electric utility infrastructure to support data communications. The Main.net PLUS solution represents a fully operational solution that connects every outlet in a subscriber's home to the electric utility backbone and includes the PLUS Network Management (NmPLUS), which provides the utility with the ability to manage and control tens of thousands of PLUS units deployed to end users.

Smart Repetition

The Main.net PLUS solution is built on the concept of using a minimal signal level instead of the strongest possible transmission signal to reach the end user. To perform this technique, which minimizes the level of potential interference, each device transmits the lowest possible signal required to reach the next device in the network. The PLUS system uses smart repetition to regenerate signals and the attenuation encountered in the electric lines to create "cells" similar to those used in mobile telephone technology, which enables the reuse of the same frequencies by nonadjacent cells without causing interference between cells. The PLUS system also uses a proprietary Point-to-Multipoint Protocol to support communications between PLUS devices. For the subscriber, the operation of the PLUS system is a transparent operation.

PLUS Architecture

The Main.net PLUS system was developed to support both North American and European utility systems. The PLUS system consists of three categories of equipment. The first category is access units, which are installed by the power utility along the power lines and form the network backbone. The second category is in-home units, which are plug-and-play modems that are installed by the end user, and the third category is network management hardware and software installed at a utility's regional control center.

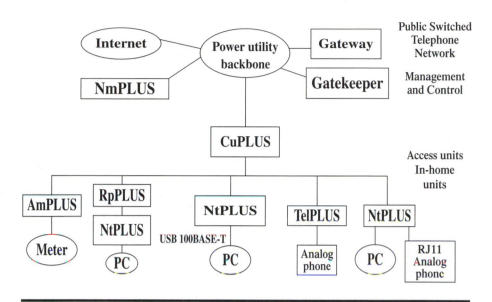

Figure 6.3   Relationship of Main.net hardware and software.

Figure 6.3 illustrates the relationship of Main.net hardware and software to the electric utility network backbone and subscriber operations.

CuPLUS

The CuPLUS represents a concentrating unit that is located in the vicinity of the low-voltage transformer serving several homes. The CuPLUS transforms information from the backbone onto the utility electric grid and from the electric grid to the backbone network. The CuPLUS communicates with home and other access units on the downlink while communicating with the NmPLUS network management system on the uplink. Several CuPLUS units are used to form a commercial system, with the backhaul from several CuPLUS units used to create a "PLC cell."

RpPLUS

The RpPLUS represents an enhanced repeater that connects an indoor PLUS unit to the network. This device enables extended transmission distances to be achieved even in noisy environments.

CtPLUS

A third component is the CtPLUS, which represents a communications transformer solution for the 110/120-V ac market in North America. The CtPLUS enables electric utilities to transmit data over medium-voltage lines that have 110/120-V ac grids.

AmrPLUS

The AmrPLUS is an automatic meter reading (AMR) unit. This device, as its name implies, enables an electric utility to automatically read a customer's meter. The AmrPLUS includes an interface that enables a series of AmrPLUS devices to be connected to AMR concentrating units.

In-Home Devices

In the lower portion of Figure 6.3 several devices to the left of the AmrPLUS were omitted to simplify the diagram. Those devices represent in-home units that provide connectivity between the electric outlets in a home and the backbone network. The NtPLUS network termination represents the basic power line modem that provides a connection between the electrical outlet to the network and a computer or peripheral device in the home. The NtPLUS includes USB and Ethernet connectors and an optional RJ11 telephony connector, the latter enabling a standard telephone to obtain a VoIP capability. The TelPLUS can be viewed as a simplified NtPLUS developed for telephony because it has only an RJ11 connector.

NmPLUS

In the upper left portion of Figure 6.3 you will note NmPLUS, which represents Main.net's network management product. NmPLUS provides an electric utility with both network monitoring and diagnostic capability, enabling the utility to manage and monitor all PLUS units from a central location. Through the use of NmPLUS, the utility can remotely enable, disable, or terminate service to a customer.

NmPLUS includes several features associated with conventional network management products. Those features include providing basic statistics, traffic counting that can be used to create billing records, the display of the topology of the components in the network, and an online display of PLUS units in the network. The PLUS system includes a self-learning capability and, as previously mentioned, has the ability to enable and disable system components, including blocking services to customers who fail to pay their bills.

Customers

Main.net has an international customer base, ranging in scope from Poland to India and the United States. In Poland, Main.net's technology was used for piloting PLC services over the power grids of Energetyka Poznanska S.A. and Lubelski Zaklady Energetyczny S.A. Energetyka Poznanska S.A. is the second largest electricity distributor in Poland; Lubelski Zaklady Energetyczny S.A. serves 500,000 customers in southeastern Poland and is the 12th largest electricity distributor in that country.

In India, Main.net Communications Ltd. provides PLC to the State Secretariat of Assam. Together with the Hyderabad-based company Apind Online and the U.S. company 22Northtech, Main.net installed a PLC system in the State Secretariat of Assam to provide Internet access through electrical outlets.

In the United States, Main.net Communications Ltd. announced a feasibility study on the electric grid of PPL Electric Utilities Corporation, a large utility headquartered in Allentown, Pennsylvania, which has 11,500 MW of generating capacity in the United States. In addition, as previously mentioned, Main.net is the first provider of BPL equipment to have its products move from a field trial to actual deployment in the United States. Currently, Main.net's worldwide customer base includes more than 40 electric utilities located in more than 15 countries. Due to the extensive number of worldwide tests and deployments of Main.net equipment, this organization is considered to represent a global market leader in PLC technology solutions.

Siemens Corporation

This large-scale electronics company headquartered in Germany manufactures a number of products designed to facilitate communications over power lines. Beginning in 2002, Efficient Networks Inc., a subsidiary of Siemens Corporation that later became a new global unit of the corporation called Siemens Subscriber Networks, released a series of power line home networking products referred to as Speed Stream. Members of the Speed Stream family include the Speed Stream Powerline USB Adapter (2501), the Speed Stream Powerline Ethernet

Adapter (2502), the Speed Stream Powerline 802.11b Access Point (2521), and the Speed Stream Powerline Wireless DSL/Cable Router (2524). In this section we will briefly discuss each member of the Speed Stream family.

Speed Stream Powerline 2501 USB Adapter

The Siemens Speed Stream Powerline 2501 USB Adapter enables users to connect a PC to a home network via a USB cable. This device is then plugged into an electrical socket to enable communications over the electrical wiring in a home or office. To prevent unauthorized reading of data, all Powerline devices support 56-bit DES encryption and comply with the HomePlug 1.0 standard to achieve 14 Mbps throughput capability.

Speed Stream Powerline 2502 Ethernet Adapter

The Speed Stream Powerline 2502 Ethernet Adapter connects any computer having an Ethernet port to the network operating over in-building electrical wiring. Similar to the Siemens' 2501, the 2502 consists of an Ethernet jack and electrical plug.

Speed Stream Powerline 2521 Wireless Access Point

The Siemens Speed Stream Powerline 2521 Wireless Access Point combines the ability to transmit and receive data over in-building electrical wiring with over-the-air wireless transmission. Concerning the latter, the Siemens 2521 product supports the IEEE 802.11b standard, which enables wireless transmission at data rates up to 11 Mbps. It should be mentioned that Siemens introduced a gateway that combines DSL access with 802.11g wireless and power line networking capability, referred to as the Speed Stream 6400. Thus, this newer product combines three popular networking technologies into one package.

Speed Stream Powerline 2524 Wireless DSL/Cable Router

The Siemens Speed Stream Powerline 2524 Wireless DSL/Cable Router represents a device that can be used to form a network hub within a home or small office. The 2524 functions as a wireless access point to support IEEE 802.11b over-the-air transmission while offering power line connectivity that turns the electrical wiring in the home or office into another transmission medium. Through the integration of four Ethernet ports, the 2524 is able to support conventional wired Ethernet connectivity, while a WAN port enables the device to be connected to either a DSL or cable modem to provide Internet connectivity for users to access the device via electrical wiring, Ethernet cabling, or over-the-air transmission.

6.2 BPL Field Trials

At the time this book was researched, most BPL operations in the United States occurred through field trials that were limited to providing service in a small neighborhood, ranging in scope from a few blocks to a suburban area where homes are clustered. Such field trials were performed as a mechanism to test the technology, to obtain familiarity with the operation of the technology and such issues as billing and maintenance, or to better understand the economic issues facing a utility prior to reaching a decision concerning an expanded rollout of the technology.

One side effect of several field trials was related to interference caused by transmission over power lines. As previously noted in this book, BPL can result in interference to high-frequency bands. During 2004, approximately 40 official complaints related to BPL interference with amateur radio operations were filed with the Federal Communications Commission (FCC). Although such complaints were considered by the FCC in its promulgation of rules that excluded BPL systems from using certain frequencies, BPL can still use certain frequencies that can adversely affect amateur radio operations. Thus, it is quite possible that the results of ongoing and future field trials could result in a modification to the list of frequencies that BPL systems cannot use.

When this book was researched, there were a number of utilities using equipment from four vendors during their field trials. Table 6.1 indicates the relationship between utilities and vendors with respect to the use of a particular vendor's equipment. Readers should note that just because a particular electric utility was using equipment from a vendor to perform a BPL field trial does not mean that the utility would continue to use equipment from that vendor if it decides to expand the scope of the field trials or conduct a major deployment of BPL technology. In fact, as you examine the use of vendor equipment, you will note several utilities are using equipment from different vendors.

Table 6.1   Utility Field Trials and Equipment Vendors

Ambient Corporation
  ConEd
  Southern Company
  Idaho Power
  Orange & Rockland
Amperion
  American Electric Power
  Alliant
  Bowling Green
  Hawaiian Electric
  Idaho Power
  PG&E
  PPL
  Progress Energy
  Southern Company
  TXU
Current Technologies
  Cinergy
  Hawaiian Electric
  Kissimmee Power Authority
  Pepco
Main.net
  Ameren
  Avista
  City of Manassas
  Dominion
  FPL
  PG&E
  PPL
  Southern Company
  TECO

Perhaps a key exception to the potential use of different vendor equipment from a field trial to a deployment is the City of Manassas. The City of Manassas used equipment from Main.net for its field trial and continued to use such equipment when it moved from a field trial environment into a citywide deployment.

In addition to the vendors listed in Table 6.1, a number of vendors can be considered to represent emerging powers in the BPL market. Some vendors manufacture chipsets that are used in other vendor products; other vendors, such as DB2, a Spain-based manufacturer, recently introduced a series of products developed for facilitating communications over power lines.

Alliant Energy

Alliant Energy Corporation is an energy holding corporation that provides customers in the Midwest with electric and natural gas services. In March 2004, Alliant Energy began a pilot BPL test in Cedar Rapids, Iowa.

Background

The Cedar Rapids pilot BPL test went live on March 30, 2004, and was expected to remain active until August or September of 2004. The pilot test involved the use of overhead power lines in an area that was previously recognized as a source of radio frequency interference (RFI) from power lines. The system used included both overhead and underground BPL connections that were connected to 2.4-GHz wireless hot spots for end-user access.

Interference Issue

During the pilot test, an amateur radio station located approximately 180 meters from a power line transporting BPL transmission immediately noticed increased interference. Alliant Energy cooperated with the American Radio Relay League (ARRL) and turned the BPL system on and off while a spectrum analyzer made measurements of the level of interference. The resulting data indicated that the BPL system resulted in signals that were significantly above the level of sensitivity of amateur radio receivers installed at the amateur radio operator's fixed location. Using equipment provided by Amperion, several notching schemes were attempted by Alliant Energy to minimize BPL interference, with limited success.

Alliant Energy terminated its pilot BPL test prematurely, shutting down the trial on June 25, 2004. According to the Alliant Energy BPL project leader, Alliant Energy was able to "accomplish the majority of its objectives" ahead of schedule because the primary purpose of the Cedar Rapids evaluation was to gain an understanding of BPL technology, including any problems that might occur. The Alliant Energy project leader also mentioned that the topology of the area presented challenges, especially with respect to the use of IEEE 802.11 wireless hot spots. In addition, the project leader suggested that Cedar Rapids might not have been the best location for a field trial. Interestingly, the Alliant Energy project leader felt that significant progress has occurred with respect to interference and that the company might take another look at BPL once the FCC places BPL rules and regulations into effect and the technology further evolves.

City of Manassas

The City of Manassas, Virginia, was the first major location in the United States to commercially offer high-speed Internet access through electric power lines. It also represents the first major location in the United States to move from a field trial into a full-scale deployment based on lessons learned from the field trial.

Pilot Project

The initial deployment of equipment to provide Internet services occurred in late 2003 and was made possible by a grant awarded through the American Public Power Association (APPA) in 2001. The APPA grant was for the initiation of a pilot project to evaluate the delivery of high-speed Internet service via the city's electrical distribution system. During the pilot project, the City of Manassas made use of its existing 30 miles of previously installed fiber-optic wiring for use as a backbone. The backbone was supplemented by BPL equipment provided by Main.net Communications, an Israel-based company, to construct an end-to-end communications network.

The Franchise Agreement

The pilot project received favorable reviews from city residents, who were able to receive high-speed Internet access by simply plugging a modem into any electrical outlet in their home. As a result of the favorable view of the pilot project, in October 2003, the Manassas City Council voted unanimously to award a ten-year franchise agreement to Communications Technology, Inc. (COMTek), a private organization, to develop Internet access citywide. The franchise agreement with COMTek represents the first full-scale commercial deployment of Internet access over power lines to occur within the United States.

Under the franchise agreement, the City of Manassas operates the fiber-optic and BPL network as "open access" and shares in the subscription revenues from the sale of broadband services. City engineers are responsible for installing and maintaining BPL and fiber-optic network components necessary to provide Internet services to city residents; COMTek, which is the network owner and operator, is responsible for acquiring BPL equipment, performing network engineering, and monitoring activities, customer support, and account billing. COMTek, in turn, uses BPL equipment from Main.net Communications.

City Infrastructure

The City of Manassas is located approximately 30 miles southwest of Washington, D.C. The city has a population of approximately 36,000 residents within the ten square miles of its boundary. Within the city are 15,000 electric meters, 12,470 volt distribution units, and 1,965 transformers, with approximately 80 percent of power lines running underground. Based on the routing of low-voltage lines from transformers to customers, there are 7.5 customers per transformer, which is approximately double that of suburban locations. The extensive underground infrastructure made the deployment of any new cable, such as fiber, to the home a daunting task, making BPL very attractive because transmission occurs over the existing power line infrastructure.

Use of Existing Fiber

Because the City of Manassas had previously installed more than 30 miles of fiber-optic cable, the fiber was used for the backhaul. This resulted in the creation of a hybrid fiber–broadband power line (HFBPL) technology; BPL was used to extend communications to homes and offices at minimal cost because it would be extremely expensive to route fiber directly to the subscriber. The fiber backhaul,

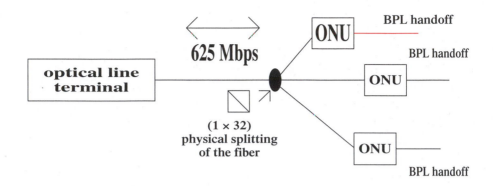

ONU Optical network Unit

Figure 6.4 Using the Ethernet passive optical network to create a hybrid fiber–broadband power line network.

although passing through much of the city, was not directly routed to distribution switches, where it would be needed to interface with BPL equipment. This is because directly connecting fiber pairs to each switch would have rapidly depleted the existing fiber network. Therefore, the City of Manassas used passive optical network (PON) technology. With PON, which was originally developed as a mechanism to support fiber to the home, a single optical fiber is "split" so that it can support up to 32 locations. Using PON technology required overcoming a second problem, because in North America the technology is primarily based on Asynchronous Transfer Mode (ATM) protocols whereas BPL is based on Ethernet. Because ATM-based PON solutions employ asymmetrical bandwidth and BPL equipment used by the city supports symmetrical transmission, the use of ATM-based PON could potentially create network bottlenecks. The solution to the PON problems resulted from the use of Fujikura Ethernet-based equipment that operates over a standard PON physical infrastructure.

Originally developed to support Japan's fiber-to-the-home (FTTH) technology, Fujikura's E-PON technology provides symmetrical bandwidth and an Ethernet connection to Main.net's BPL equipment. Figure 6.4 illustrates the use of PON technology to form a HFBPL infrastructure.

In examining Figure 6.4, note that each optical network unit provides an interface for a BPL handoff. The BPL handoff is actually located at one of the numerous distribution switches throughout the City of Manassas, with the placement of Ethernet data carried on the backhaul onto the electrical grid performed by an injector.

Injector Operation

An injector is installed at each of the 600-amp distribution switches. Each injector injects Ethernet data onto the electrical grid. From each distribution switch, between 50 and 75 customers are able to receive 500-kbps service with existing Main.net equipment; however, it is anticipated that next-generation equipment will enable data rates up to 1.5 Mbps to be achieved.

Repeater Utilization

Between the distribution switches and subscribers, repeaters are installed at step-down street-level transformers. In addition, depending on the distance between the subscriber's home and the serving transformer, a repeater may also be installed in a home's electrical meter base.

Construction Cost

With an average of 7.5 customers per transformer, it cost from $400 to $600 per customer to construct the BPL network from the distribution switch into the subscriber's home (assuming a 20 percent penetration rate). The cost includes providing the home user with a modem that can be used at any electrical outlet in the city where service is offered. The cost is rapidly regained from subscription fees of $28.95 per month for residential service and $39.95 per month for commercial service. Once next-generation BPL equipment is installed that can provide higher data rates, COMTek may offer Manassas customers tiered services, so that customers will pay different rates to obtain higher data transfer capabilities.

Technical Issues

As indicated several times in this book, BPL faces a number of technical issues. One key issue is interference. Because BPL in the City of Manassas is primarily on underground electrical cables, the major area of interference results from data signals being broadcast over transformers if the signals are allowed to traverse the transformer switches directly. However, because Main.net Communications equipment is based on the concept of smart repetition, in which signals are repeated only if necessary, low signal levels result in minimal interference that does not adversely affect other radio operations at reasonable distances from overhead BPL locations. Unfortunately, because of the unique characteristics of Main.net technology, all components of its system, including its NtPLUS home modems, are proprietary. This means that both residential and commercial customers who wish to subscribe to the citywide BPL service do not have the ability to select a HomePlug-compatible device available from many other vendors. Instead, they must use an NtPLUS modem from Main.net.

Current Broadband

Current Communications has partnerships with Cinergy Corp., an electric utility, to offer broadband communications to electric customers in the greater Cincinnati, Ohio, and northern Kentucky areas. According to newspaper reports, although Cinergy and Current will not disclose how many homes have signed up for the service, the intention of both organizations is to have a service capability that will enable up to 250,000 homes to subscribe to BPL service within three years.

Service Offerings

Current Broadband offers subscribers three levels of service under the designator's "Current Quick Access," "Current Premier Access," and "Current." Current Quick Access provides a maximum data rate of 1 Mbps, whereas Current Premier Access provides a maximum data rate of 3 Mbps. The Current service also provides a maximum data rate of 3 Mbps; however, it includes a fixed IP address as well as the ability for subscribers to host their own Web sites. Table 6.2 provides a comparison of the three Current Communications product offerings.

Discount Rates

At the time this book was written, Current Communications was offering discount rates for customers who subscribed to a two-year period. For Current Quick Access, the two-year rate was $26.95 per month, and the monthly rates for Current Premier Access and Current were $34.95 and $44.95, respectively.

Table 6.2 Current Broadband Product Offerings

Product	Maximum Data Rate	Web E-Mail Addresses	Personal Web Site Storage	Static IP Address	Monthly Cost
Current Quick Access	1 Mbps	5	5 MB	No	$29.95
Current Premier Access	3 Mbps	5	5 MB	No	$39.93
Current	3 Mbps	10	10 MB	Yes	$49.95

Progress Energy

This Raleigh, North Carolina, based utility holding company conducted a two-phase BPL field trial during 2004. Progress Energy teamed up with EarthLink, the second largest ISP in the United States, to test broadband Internet service in several Wake County, North Carolina, neighborhoods via power lines. The two-phase field trial represented the first time that high-speed Internet service was commercially offered via power lines in North Carolina. However, in contrast to the City of Manassas, the North Carolina trials at this time do not represent a full BPL deployment.

Overview

During the two-phase field trial, Progress Energy used technology developed by Amperion, Inc., of Massachusetts. Amperion equipment in this field trial used a combination of fiber optics and power lines to transmit data to a relay point in certain neighborhoods. At the relay point, IEEE 802.11 equipment was used to service subscribers via wireless transmission. Approximately 500 homes had the opportunity to participate in the field trial, with service costing $19.95 per month for the first three months and $39.95 per month thereafter. As part of the trial, Progress Energy also upgraded some electric meters so they could be remotely monitored through the BPL system. In addition, monitoring equipment enabled outages in the area of the field trial to be tracked, and line and service personnel working within the area of the field trial could use the BPL technology to access the company intranet and send and receive e-mail.

Technology

Progress Energy's field trial was based on technology provided by Amperion, Inc. The phase 1 trial performed during 2002 included both residential and commercial customers and demonstrated the technical feasibility of BPL. The company's phase 2 trial, which commenced in early 2004, was designed to test the economics associated with BPL. Under the phase 2 trial, Progress Energy used IEEE 802.11b connections at the service injection point for medium-voltage lines and at the subscriber drop as indicated in Figure 6.5.

In examining the architecture shown in Figure 6.5, note that, similar to other field trials performed using different vendor equipment, injectors,

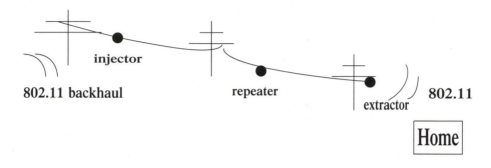

Figure 6.5 The Progress Energy field trial used WiFi 802.11b wireless connections at two locations.

repeaters, and extractors are used. The repeaters boost the signal while line-mounted extractors, which can be powered through induction, include an internal WiFi antenna. In other areas of the field trial, WiFi antennas were hidden inside light poles.

What distinguishes the Progress Energy field trial from other field trials is its use of wireless LAN technology. Although most field trials either use or have the ability to optionally support IEEE 802.11 wireless LAN technology to bypass the use of low-voltage lines to the home or office, the Progress Energy field trial also supports the use of wireless LAN technology from medium-voltage lines onto the backhaul. Thus, the Progress Energy field trial uses wireless LAN technology as a mechanism to connect both to customers as well as to the backhaul.

Architecture

The architecture used in the Progress Energy field trial can be thought of as a BPL to the neighborhood or a final over-the-air solution, because WiFi provides the connection into the home and small business. During the field trial, it was determined that line-mounted injectors and extractors could be installed in approximately three minutes once personnel were positioned at an appropriate location. The BPL facility used in the field trial provided a 20-Mbps channel of combined upstream and downstream capability per injector; however, the 802.11b equipment used in the trial had a maximum data rate of 11 Mbps, and the obtainable data rate decreased as the distance between the WiFi connection point and subscriber increased. The backhaul portion of the BPL network was constructed through the use of subrate DS-3 (T3) transmission facilities that were also accessed via WiFi communications.

The Progress Energy field trials lasted six months and cost approximately $500,000 through August 2004, when it was terminated. During the six-month phase 2 trial, approximately 400 customers subscribed to the service. Although Progress Energy has stated that it has no immediate plans to offer high-speed Internet service via its power lines, according to the company, the trial was successful and they "gathered valuable information about broadband over power lines and its potential." According to a Progress Energy spokesperson, the company is examining whether it should develop a business model to plan a large-scale deployment of BPL.

Chapter 7

Evolving Standards and Organizing Bodies

In writing a book on broadband over power lines (BPL), which represents an evolving technology, I cannot fail to discuss two emerging Institute of Electrical and Electronics Engineers (IEEE) standards as well as their relationship to other standards and the efforts of several organizing bodies. Thus, the initial focus of this chapter is on the work of the IEEE to standardize BPL technologies. Then we will turn our attention to a brief discussion of other standards and the work of several organizing bodies that can be expected to affect the manufacture of BPL equipment and their operational capability.

7.1 IEEE Standards

This section highlights the emerging role of the IEEE in the BPL standardization process. We will discuss two IEEE working groups and the results of a series of meetings that occurred during 2004 and 2005.

Overview

The efforts of the IEEE date back to April 26, 2004, when a Project Authorization Request (PAR) was filed and assigned Project Number 1675. Now referred to as P1675, the formal title of the PAR is "Standard for Broadband over Power Line Hardware." The referenced PAR is sponsored by the IEEE Power Engineering Society Power System Communications Committee (PES/PSCC). When completed, the IEEE P1675 effort will provide electric utilities with a comprehensive standard for installing hardware on both overhead and underground power distribution lines, which form the infrastructure for constructing and operating BPL systems. The P1675 project represents one of two related standards efforts undertaken by the IEEE. The second effort, which was assigned project number P1775, is more formally referred to as "Powerline Communications Equipment — Electromagnetic Compatibility (EMC) Requirements — Testing and Measurement Methods." The PAR for this project was approved on May 10, 2005, and is also sponsored by the IEEE PES/PSCC. According to the P1775 PAR, the proposed standard is part of a planned IEEE series of BPL standards that will cover such major aspects of BPL technology as equipment operation, interoperability, safety, EMC, media, and even education. Because of the recent approval of the P1775 project and its orientation toward developing EMC criteria and test and measurement procedures for BPL, which are due for submission for an initial ballot in 2007, this chapter will focus primarily on the P1675 project.

P1675

The IEEE P1675 PAR form indicates that the project will result in the development of a new standard. The scope of the project is to provide for the development of standards for the testing and verification of commonly used hardware associated with the installation of BPL systems. Such hardware will include injection and extraction couplers and enclosures; however, the project is not designed to cover repeaters, node hardware, data transmission protocols, and other aspects of BPL that are related to the internal workings of the technology.

The stated purpose of the P1675 effort is to develop standards that will allow both utilities and equipment vendors to acquire and manufacture standardized equipment that enables use of such equipment without worrying about safety and usability. Because the standard will also define who will perform equipment installation, such as linemen installing couplers on medium-voltage lines, another goal of the standard is to define the boundaries of the working areas associated with qualified linemen and other personnel.

According to the original PAR form, the completion date of the projects submittal is May 1, 2006. Once completed, the IEEE P1675 standard will provide electric utilities with a common method for installing hardware on both underground and overhead power distribution lines as well as define the installation requirements for the protection of personnel who install BPL equipment. Through the use of this evolving standard, electric utilities will be able to assess both the performance and safety of devices intended to be used on medium- and low-voltage lines that provide the infrastructure for a BPL capability. In addition, the P1675 standard is expected to tackle such issues as how to keep installed equipment operating and prevent such equipment from interfering with the primary purpose of the power system, which is the delivery of electricity.

Focus

The focus of the evolving P1675 standard is not on specific types of hardware nor on a particular design. Instead, the standard is focused on determining the specifications of equipment that enable devices to be installed so that they operate correctly under different environments and do not adversely affect other devices nor result in a hazard to utility personnel or customers. For example, a coupler will be required to have a particular surge withstand capability and certain environmental characteristics, such as the ability to withstand ultraviolet rays, vibration, and resistance to the elements. In addition, the standard can be expected to define a basic insulation level (BIL), which, in effect, will denote a minimum level of device protection as well as tests required to verify the operation of equipment. The standard will include an installation section that will supplement the hardware portion, denoting the specifications required to properly install hardware. For example, for overhead power lines, the installation section could include information concerning the best method for grounding and bonding equipment. In an underground environment, where transformers are mounted on concrete pads, it becomes very important to keep people away from harmful voltages; thus, the standard may define the housing required for transformers as well as the safest method for coupling the primary and secondary voltages inside a pad-mounted transformer.

Interference Issues and PAR 1775

Perhaps the key omission in the P1675 PAR regards the issue of electromagnetic interference. As previously noted in this book, BPL transmission over medium- and low-voltage lines results in the use of orthogonal frequency division multiplexing (OFDM), under which data signals occur on carriers spread over a range of frequencies. Some of those frequencies originally interfered with various ham radio operations, baby monitors, and other types of radio operations. This interference resulted in the U.S. Federal Communications Commission (FCC) issuing a list of frequencies that BPL equipment must avoid.

Electromagnetic interference represents a valid issue that is not addressed by the P1675 emerging standard. Perhaps because of this, another PAR 1775 was initiated. As mentioned earlier in this chapter, P1775 is more formally referred to as "Powerline Communications Equipment — Electromagnetic Compatibility (EMC) Requirements — Testing and Measurement Methods" and is oriented toward EMC issues.

P1675 Group Meetings

Similar to other IEEE projects intended to develop standards, the P1675 effort involves a series of project meetings that can be expected to result in the development of a standard. In this section we turn our attention to several IEEE BPL study group meetings that transpired during the middle to latter portion of 2004 and early 2005. We will discuss the goals of each meeting, different presentations that occurred during those meetings, if applicable, and any deliverables and schedules that resulted from the meetings. This information will provide readers with an understanding of the evolution of the P1675 standard.

June 2004 Meeting

The first meeting came about because of a "Call for Interest in Standards on Broadband over Power Lines." Occurring June 7, 2004, in Denver, Colorado, the meeting focused on understanding how the standardization of BPL technologies will accelerate development in this field as well as determining what resources are available from industry and IEEE societies to develop applicable standards.

The June 7, 2004, meeting began at 10 a.m. and was adjourned at 3 p.m. After an introduction and overview of the goals for the meeting, the IEEE PES provided a presentation, followed by a series of approximately 15-minute presentations from electric utilities, vendors, the American Radio Relay League (ARRL), and the HomePlug Powerline Alliance.

Presentations

One of the more interesting presentations at the June 7, 2004, meeting was given by Amperion, concerning regulatory considerations. The speaker noted that power line vendors must comply with the FCC's radio frequency (RF) emission standards, which is not exactly new news. However, the presenter then went on to mention that BPL vendors must determine the method to proportion fees between regulated utilities and service providers, which, although not a standards issue in my opinion, represents an issue that needs to be addressed by any electric utility that plans to offer a BPL capability. Although an electric utility needs to be compensated for the use of its infrastructure and the installation and maintenance of BPL equipment, its fees should not reach a point such that they discourage the deployment of a communications capability. To the best of my knowledge, this portion of the presentation was the only time this important issue was raised.

A second informative presentation at the June 2004 meeting was from a member of AT&T Laboratories. This presenter discussed interference and its mitigation in BPL from the perspective of a service provider. Topics covered during this presentation included the relevant Code of Federal Regulations (Part 15, Title 47), which provides FCC rules and regulations, the causes of interference, and the potential of adaptive interference management. The June meeting concluded with a discussion concerning the next steps that should occur. Several key points were made. One point was that the utilities would play a major role in the standardization process, and it would be necessary to coordinate the effort with the HomePlug Powerline Alliance and other standards organizations. In addition, a potential layout of the standardization effort was proposed. That suggestion indicated that the standards effort should focus on four areas: (1) construction, to include safety and reliability, (2) media, (3) the physical layer, to include the interface to the media, and (4) Media Access Control (MAC).

One of the key points raised during the discussion of the next steps to take was that the IEEE should not take too long to develop BPL standards. A delay in the standardization process could stall the marketplace. At the conclusion of this preliminary meeting, a follow-up two-day meeting was proposed for July 20–21 at the IEEE operations center, located in Piscataway, New Jersey.

July 2004 Meeting

The second IEEE BPL meeting occurred July 20, 2004, at the IEEE headquarters in Piscataway, New Jersey. Although it was originally planned as a two-day meeting, the meeting was conducted during a single day. This meeting involved the formulation of a mission statement, position statements on relevant BPL issues, a discussion of whether to produce a white paper on BPL, a series of presentations, and a discussion of future steps. The meeting concluded with the selection of October 13, 2004, as the date of the next meeting, which would also be held at the IEEE headquarters in Piscataway, New Jersey.

Mission Statement

One of the more important outcomes of the second study group meeting was the development and agreement of attendees to the formulation of a mission statement. The mission statement can be viewed as a road map for the development of BPL standards and is as follows:

> The mission of this group is to draft BoPL R&D scenarios and establish a scope and produce recommendations for standardization at IEEE-SA and relevant positioning with respect to other industry and standards bodies.

Position Statements

After formulating the mission statement, attendees devoted a good portion of the second IEEE BPL meeting to position statements and presentations from utilities, other IEEE groups, the ARRL, and other standardization bodies, such as the HomePlug Powerline Alliance. Both position statements and presentations noted that HomePlug 1.0 technology is already operational on in-home electrical wiring and needs to be taken into consideration. The HomePlug presenter noted that HomePlug AV is positioned for launch in the near future and that it is important to collaborate with other organizations to build the industry. The ARRL presentation focused on the issue of interference, whereas a Telcordia Technologies presentation mentioned that safety issues need to be examined and suggested that existing standards and their revision process could be used to address new issues. Similarly, a presentation from a representative of Ambient also addressed existing safety standards and recommended the adoption of sections of those standards, such as insulators and surge withstand levels.

October 2004 Meeting

The IEEE BPL study group's next meeting occurred October 13, 2004. This meeting had four primary goals. The first two goals were the development of a set of recommendations for the IEEE and the compilation of a report that would include study group recommendations. The third goal of the meeting was to "move forward with a unified effort," and the last goal recognized the need for a road map.

During the October 2004 meeting, the role of the IEEE Standards Association (IEEE-SA) was discussed. This discussion included the requirement of a sponsor or sponsors for a standards project and the fact that a defined scope of work would be required before an IEEE-SA could commit resources. The IEEE-SA president-elect provided information about trade issues and the need to avoid a discussion concerning costs, industry alliances, and similar types of information when developing the BPL standards.

Collaboration

Because the development of BPL standards can affect many other areas of communications, collaboration within the IEEE society structure was recognized. This recognition resulted in a discussion concerning the need to involve the Power Engineering Society's PSCC, Power System Relaying Committee (PSRC), and Transmission and Distribution (T&D) Committee, the IEEE Communications Society (ComSoc), the EMC Society, and the Antennas and Propagation (AP) Society. As work progressed on BPL, it was determined that the appropriate areas within the IEEE structure would require various degrees of collaboration.

White Paper Discussion

To facilitate the road map goal of the meeting, attendees discussed the development of a white paper. It was agreed that a white paper would be necessary to provide a road map for the standardization process. Furthermore, it was determined that the BPL Web site hosted by the IEEE should be easier to locate on the IEEE-SA Web site. At the time of this writing, the URL for the BPL Web site was http://grouper.ieee.org/groups/bpl/.

Presentations

During the October 2004 meeting, there were several presentations from industry, the IEEE PES/PSCC, and a representative of the Home-Plug Powerline Alliance.

ARRL Presentation — The presenter from the ARRL discussed the need for protection from harmful interference and the fact that BPL standards need to address EMC issues. Concerning the latter, the ARRL presenter stressed the fact that the IEEE EMC Society should be directly involved in the EMC aspects of any IEEE standard that would affect the BPL industry. The ARRL presenter felt that the EMC components of a standard would need to balance the requirement of the BPL industry to have a workable environment for the manufacture and marketing of BPL technology against the need for licensed radio services to operate within a home environment without harmful interference becoming an issue.

Panasonic Presentation — A second presentation was from Panasonic Corporation. The presenter noted that the electric power meter could be considered to represent a logical candidate for use as a residential gateway. In this presentation, the Panasonic presenter also mentioned that quality of service (QoS) needs to be preserved in the home, with home networking having priority regardless of what additional applications use the remaining bandwidth.

Panasonic views BPL as being divided into three levels. An economy level would provide data rates from 64 to 384 kbps and be oriented toward the mass market. A business class would provide a data rate of 1 to 2 Mbps and be oriented toward selected markets, whereas a premium class of BPL service would operate at 10 Mbps and provide a QoS capability. The premium class of BPL service, according to Panasonic, would not be practical for the mass market.

According to the Panasonic presenter, because approximately 90 percent of consumers do not network devices, BPL provides the ability to achieve a mass market for products greater than other access technologies. In concluding his presentation, the Panasonic representative proposed that a report should be completed for the IEEE-SA that would include the BPL MAC and physical layers, interference management, and safety standards.

IEEE PES/PSCC Presentation — During the October 2004 meeting, a representative of the IEEE PES/PSCC identified four key areas that are needed for standardization. Those areas are construction and safety, media, physical and MAC layer, and emissions.

PPL Presentation — A representative of the electric utility PPL discussed the goals of an electric utility and interfacing BPL with the electric grid. The presenter mentioned that broadband is treated as a foreign attachment and needs to acquire its own right of way. The PPL presenter went on to discuss integration standards, equipment standards, product testing, installation issues, and service restoration and maintenance.

HomePlug Powerline Alliance Presentation — The last presentation at the October 2004 meeting was provided by a representative of the HomePlug Powerline Alliance. The presenter discussed co-existence as a level of service and noted that "standards provide the fundamental framework within which all of these competing services can co-exist without interfering with one another." The HomePlug Powerline Alliance presenter also discussed the emerging HomePlug AV standard, noting that it is in the final stage of its development.

Deliverables

During the October 2004 study group meeting, it was determined that the group would pursue the development of two deliverables. The first would be a recommendation to the IEEE outlining the work to be pursued. The second deliverable would be a white paper on BPL. Prior to adjournment, the next meeting of the IEEE BPL study group was scheduled for January 14, 2005, in San Diego, California, in conjunction with the IEEE PES/PSRC meeting.

January 2005 Meeting

The January 14, 2005, IEEE BPL study group meeting took place at the Marriot Del Mar hotel in San Diego, California. This meeting began with opening remarks concerning the prior meeting of October 14, 2004, and the goals for the current meeting. This was followed by a discussion of recommendations that included such areas as emissions, RF characterization of power lines, standards work being performed by other bodies, RF measurements, and standards for obtaining immunity from interference. Following that discussion, a Q&A portion of the meeting brought out the opinion that the IEEE should not be assuming the role of regulatory bodies; participants discussed the need to balance the interests of licenses and industry.

Presentations

Similar to prior study group meetings, the January 2005 meeting included a series of presentations, some of which included several recommendations. There were presentations on behalf of the Universal Powerline Association and the IEEE ComSoc, as well as a discussion of safety issues that covered both overhead and underground medium-voltage hardware. A discussion of stakeholder positions, including service management, technology suppliers, and security, privacy, and authentication occurred. Concerning the latter, this resulted in the question of whether security should cover vandalism due to interference knocking out a BPL network. Other presentations covered compatibility with wireless services, consumer use, business use, and potential service architectures.

Recommendations

The January 2005 study group meeting can be viewed as a significant milestone. At this meeting, the participants agreed to divide the group into a series of subgroups to address issues for recommendations and white papers. The subgroups would be responsible for developing their own PARs, as applicable, and would then finalize submissions for the white papers that are relevant to each subgroup. Subgroups were established for emissions/ECM, MAC, and PHY layers, safety and construction, media, and education. The study group had planned for the next meeting to occur concurrently with the UTC Telecom 2005 Conference May 25 and May 26, 2005, at the Long Beach Conference Center, in Long Beach, California. Unfortunately, that meeting was canceled in conjunction with UTC Telecom 2005. However, two of the three working groups were not affected and met. The IEEE P1675 working group met on May 25 and the IEEE P1775 working group met on May 26. At the time this book was prepared, the results of those two meetings had not been posted on the IEEE BPL Web site, nor was there any mention on the Web site of a subsequent meeting. However, because the goal of the P1675 working group was to complete its standardization effort within 18 months, it is quite possible that one or more additional meetings were expected to occur during 2005 to complete the project by 2006. In fact, the United Power Line Council (UPLC) Conference meeting that was scheduled for September 2005 in Dallas was selected as the venue for the second IEEE P1675 meeting of 2005. The interesting fact about this meeting is that it was listed on the UPLC Web site prior to its inclusion on the IEEE BPL Web site.

Current Status of the Standard

To determine the current status of the IEEE P1675 standard, I contacted the chairperson of the BPL standards working group. Table 7.1 contains the draft outline of the IEEE BPL standard. Note that the draft standard is divided into two parts: hardware and installation. The hardware portion of the standard covers BPL hardware characteristics on overhead and underground medium-voltage and low-voltage transmission lines as well as the testing of such hardware. In comparison, the installation portion of the standard covers both overhead and underground equipment, including security, maintainability, equipment clearances, bonding and grounding, placement of equipment, and mounting techniques.

Now that we have an appreciation for the efforts of the IEEE, let's turn our attention to both de facto and de jour standards that can have a bearing on BPL operations on a worldwide basis.

7.2 Other Standards and Organizations to Note

In this section we turn our attention to both existing and emerging standards that can have a bearing on the operation of BPL as well as other organizations whose efforts can be expected to promote BPL standards. We discussed the three HomePlug Powerline Alliance specifications in Chapter 5. This organization and its existing and evolving standards provide a good starting point for both a discussion of other standards as well as a review of existing and evolving HomePlug Powerline Alliance specifications. Thus, we begin this section by reviewing current and evolving HomePlug Powerline Alliance specifications.

HomePlug Powerline Alliance Specifications

The HomePlug Powerline Alliance represents a group or alliance of organizations involved in each level of the value chain, from services and content to retail, hardware, software, silicon, and technology. The mission of the HomePlug Powerline Alliance is to enable and promote the rapid availability, adoption, and implementation of cost-effective, interoperable, and standards-based home power line networks and products.

Table 7.1   BPL Standard P1675 — Draft Outline

Overhead Devices
MV*
    Couplers
        Capacitive (shunt)
        Inductive (series)
        Other line hardware
            Fused Cutouts
LV**
    Couplers
        Capacitive (shunt)
        Inductive (series)
        Other line hardware
            Fuses

Underground Devices
MV
    Couplers
        Capacitive (shunt)
        Inductive (series)
        Other line hardware
            Fuses
LV
    Couplers
        Capacitive (shunt)
        Inductive (series)
        Other line hardware
            Fuses

Information to be covered in the coupler section of the standard:
Couplers
BIL
Surge-withstand characteristics
Weather resistance
Environmental
    Temperature range
    Humidity
    Waterproof
    Shock/vibration
    Dielectric
    Contamination
    Mounting
    LV[c]
Compatibility with existing hardware

Table 7.1   BPL Standard P1675 — Draft Outline (continued)

Corona
   Short circuit current
   Labeling/nameplate information
   Mechanical
      Weight
      Size
      Spacing
      Attachment
      Failure mode
      Flammability
      Liquid versus dry
Grounding
Testing
   Impulse withstand
   Vibration
   Environmental
      Sunlight Resistance
   Short circuit withstand
   RIV
   HV[d] tests
   Power quality
   Impact resistance

Information to be covered under the installation section of the standard:

Installation
Overhead installations
   Security and tamper resistance
   Maintainability
   Equipment clearances
   Grounding and bonding
   Placement of equipment
   Mounting techniques
Underground installations
   Security and tamper resistance
   Maintainability
   Equipment clearances
   Grounding and bonding
   Placement of equipment
   Mounting Techniques
   Underground elbow — IEEE 386 and P1215

[a] Medium voltage.
[b] Low voltage.
[c]
[d] High voltage.

HomePlug 1.0

The HomePlug Powerline Alliance's first specification, referred to as the HomePlug 1.0 specification, defines a combination of an 84-channel OFDM, forward-error correction, interleaving, and automatic repeat request (ARQ) to obtain a data transmission rate of 14 Mbps over electrical wiring. This specification was released in June 2001, approximately 15 months after the HomePlug Powerline Alliance was founded.

HomePlug AV

The release of the HomePlug Powerline Alliance 1.0 specification was followed in February 2003 with commencement of work on a high-speed version of the technology that would enable data rates up to 200 Mbps. This effort is referred to as HomePlug AV, with the designator "AV" referencing "audio-visual" as well as the capability of the evolving specification to transport high-definition television (HDTV) signals over the electrical wiring in a home. The HomePlug AV specification represents a mechanism to extend the data transmission rate over electrical wiring to 200 Mbps while providing backward compatibility with equipment operating in compliance with the HomePlug 1.0 specification.

The HomePlug AV specification is expected to result in the availability of hardware that will enable support for many applications that require additional data transmission beyond the 14 Mbps of HomePlug 1.0-compatible products, as well as all versions of WiFi, to include the evolving 802.11n standard. The finalization of the HomePlug AV specification, which is expected to occur during 2006, will enable both standard-definition television (SDTV) and HDTV, including IPTV, to be supported. In addition, higher speed Internet sharing via support of data streams received via ADSL2 and VDSL can also be supported. To achieve this capability, the emerging HomePlug Powerline Alliance AV specification supports 917 usable OFDM carriers whereas the prior HomePlug 1.0 specification is limited to the support of 84 OFDM carriers. This enables the AV specification to achieve a 200-Mbps channel rate and a 150-Mbps information rate. Other physical layer characteristics that can be expected to be included in the HomePlug Powerline Alliance AV specification include a power line cycle adaptation capability, which enables adaptation to time-variant line cycle noise; a robo mode, which enables broadcast communications at 5 Mbps and 10 Mbps; a point-to-point channel adaptation capability, which uses a modulation density of BPSK to 1024 QAM per tone; and the use of forward-error correction to include turbo codes and convolution codes to minimize bit errors. At the MAC layer, the HomePlug Powerline Alliance AV specification defines the use of TDMA channel access with contention-free and contention-based CSMA/CA periods. In addition, the emerging AV specification employs a basic service set (BSS) type of architecture with a central coordinator similar to the wireless LAN BSS that uses an access point. Under the emerging AV specification, all new stations joining the BSS are both associated and authenticated. Perhaps taking a page from conventional Token Ring LANs, the emerging AV specification uses a beacon-based design. The ac line cycle functions as a synchronization for the beacon generation process, with CSMA allocations occurring during each beacon period. This results in the ability of the AV specification to provide QoS guarantees, including bandwidth reservation and tight control of latency and jitter. Because the emerging AV specification is currently more than 300 pages in length and growing, it represents a comprehensive document that will need to be carefully considered by the IEEE P1675 working group when developing standards.

HomePlug BPL

In addition to considering the HomePlug 1.0 and emerging HomePlug AV specifications, the IEEE P1675 working group will need to consider a third evolving HomePlug Powerline Alliance specification. That specification is currently referred to as HomePlug BPL and is oriented toward defining the operation of equipment on the infrastructure of electric utilities. Because work recently commenced on developing this new specification, it will more than likely be well into the next year or perhaps longer until this specification is completed.

Compatibility

The previously mentioned troika of existing and evolving HomePlug Powerline Alliance specifications are designed to interoperate with one another. To accomplish interoperability, the evolving AV and BPL specifications include co-existence modes that provide backward compatibility with HomePlug 1.0 devices and forward compatibility with devices that will comply with the eventual issuance of the two more recent specifications. Because several million devices are currently installed that operate according to the HomePlug Powerline Alliance 1.0 specification, I believe that the emerging IEEE standard will need to take into consideration all three of the previously mentioned Home-Plug specifications.

Other Organizing Bodies

In addition to the HomePlug Powerline Alliance, three other organizing bodies are involved in BPL with respect to operations occurring over the electric utility infrastructure. Those organizing bodies are the UPLC, the Power Line Communications Association (PLCA), and the Electric Power Research Institute (EPRI).

United Power Line Council

The United Power Line Council is a division of the United Telecom Council. The UPLC was specifically created to address power line communications and represents an alliance of electric utilities and technology companies working together to promote the development of BPL such that it helps both utilities and their partners. UPLC's efforts are focused in four areas. Those areas include business opportunities, regulatory and legislative advocacy, technical operability, and utility applications.

As part of its mission to provide both members and nonmembers with comprehensive data about BPL, the UPLC distributes a monthly newsletter called *Powerline*. Each issue of the *Powerline* newsletter reports on significant business, technical, and regulatory developments related to BPL. The UPLC also issues periodic information bulletins that are distributed only to members. Such members-only *BPL Information Bulletins* provide timely data concerning many aspects of BPL, ranging from legal to technical issues.

In addition to the previously mentioned publications, the UPLC conducts conferences to include exhibits of BPL equipment. When this book was prepared, the UPLC Annual Conference had just occurred at the Omni Dallas Park West Hotel from September 11 through September 14, 2005. What was extremely noteworthy about this conference was the fact that the UPLC Web site promoting the conference indicated that the IEEE BPL study group was also meeting there from September 14 through September 16, 2005.

The UPLC Annual Conference included approximately 20 sessions. The following are just some of the topics that were covered: "What the Heck Is Going On with BPL Standards," "The Best Way to Get

Going with BPL," "Another Pilot vs. Phased Implementations," "New Commercial BPL Opportunities," "BPL and the Intelligent Grid: Future Opportunities for Internal Utility Applications," "BPL Innovations from Around the World," and "The Great BPL Chip Debate — The Quest for Interoperability."

Electric Power Research Institute

The Electric Power Research Institute was founded in 1973 as an independent, nonprofit center for public interest energy and environment research. The EPRI is tasked with addressing the needs of society related to energy and the environment. This organization operates out of two major locations in the United States: Palo Alto, California, and Charlotte, North Carolina. EPRI's members represent more than 90 percent of the electricity generated in the United States.

Since 2003, the EPRI has been active in examining BPL emissions and electromagnetic compatibility issues. During the previously mentioned time period, the EPRI issued several reports concerning BPL. In 2003, the EPRI released a report titled, "The Race for Broadband Communications on Power Lines: Communications on Power Lines." This report compared chip sets and regulatory and economic issues concerning high-speed long distance communications over power lines. In 2004 the EPRI published a report titled, "Characterizing Electromagnetic Compatibility of Broadband over Powerline (BPL) Technologies." This report contained a comprehensive assessment of the EMC characteristics of different BPL technologies, discussing the effect of RF emissions, normal power system disturbances, and the operation of such devices as couplers and repeaters that have to be installed on distribution lines. This report is focused on two main areas: high-frequency signal analysis and evaluation of the compatibility of BPL hardware when subjected to power quality disturbances. Key results of the report concerning high-frequency signal analysis were shared with other standards organizations, including the IEEE P1675 working group.

A second BPL-related report published by the EPRI in May 2005 was titled, "Broadband over Power Lines (BPL): FCC Emissions Compliance Guidelines." This EPRI technical report presented an overview of FCC regulations limiting RF emissions from BPL equipment and then provided readers with guidance and details concerning the testing necessary to demonstrate compliance with the radiated emission limits specified by FCC regulations.

Because FCC regulations concerning RF emissions from BPL equipment are confusing, due to their use of "excluding frequencies" instead of listing allowable frequencies and because they lack clarity concerning recommended or required emission levels, the EPRI report translates the FCC's report and order concerning BPL into terms that are better understood by utility personnel and EMC specialists. Included in the EPRI report is a series of worksheets that enable readers to quickly format data into reports that satisfy FCC requirements. Because violations of FCC emission limits could subject electric utilities operating BPL systems to potential fines as well as costly modifications to their communications infrastructure and an interruption of service during the period required to perform the modifications, this EPRI report minimizes RF emission risk factors associated with the installation of BPL systems.

Summary

The two organizing bodies involved in BPL are actively working on various issues that affect the ability of electric utilities to provide an overlay communications infrastructure on existing power lines. The rapid evolution of BPL from a concept a few years ago to pilot projects and a deployment in the City of Manassas has resulted in each organization working on some aspect of BPL, either independently or in association with other standards bodies. Thus, readers are encouraged to check the Web sites of each of the previously mentioned organizing bodies as well as the IEEE P1675 working group Web site for the latest information concerning the efforts of each organization. To facilitate Web site checking, I have listed in Table 7.2 the Web site URL for each of the previously mentioned organizing bodies as well as for the IEEE P1675 working group and the HomePlug Powerline Alliance.

Table 7.2   Web Site URLs of Organizations Involved in BPL

Organization	Web Site URL
IEEE P1675 Working Group	http://grouper.ieee.org/groups/bop
HomePlug Powerline Alliance	http://www.homeplug.org
United Power Line Council	http://www.uplc.org
Electric Power Research Institute	http://www.epri.com

Chapter 8

The Future of BPL

Once in a while an author will go out on a limb and predict the future. In concluding this book on an evolving network technology, I would be remiss if I did not take out my old and trusty crystal ball and examine the potential future of broadband over power lines (BPL). In this chapter we will first focus our attention on the economic, technological, and regulatory issues that represent challenges to the deployment of BPL. Once this is accomplished, I will use the preceding information as a basis for predicting the future of BPL in certain economic, technological, and regulatory environments. Although making such predictions puts me out on a limb, hopefully, the limb will be sturdy enough that the prediction will not collapse onto the ground.

8.1 Challenges Facing BPL

In this section we will discuss challenges to the movement of BPL field trials and pilot tests to full-scale deployment by electric utilities. As noted, those challenges fall into economic, technological, and regulatory categories. By understanding the challenges faced by utilities, we can better understand what needs to be accomplished to provide a realistic basis for the widespread deployment of the technology. Thus, let's first turn our attention to the challenges facing BPL. Those challenges commonly reside in the areas of economics, technology, and regulation at the federal, state, and municipal levels.

Economics

Currently, the economic model for a major BPL deployment has yet to be verified. Although the City of Manassas represents the first major deployment of BPL to occur in North America, one must note that although the deployment occurred under contract with a commercial organization, the contract was issued by a municipality. Normally, municipalities are not concerned with obtaining a return on investment (ROI) nor are they subject to municipal fees and taxes that commercial organizations need to consider. Thus, the economics that allowed the City of Manassas to justify deployment of BPL are probably not applicable to investor-owned electric utilities that, although regulated, need to answer to their stockholders if they fail to obtain a given rate of return that causes their share price to fall or their dividend to be reduced.

We can obtain an indication of the viability of BPL from an economic perspective by considering the results of several field trials. The results of a series of BPL field trials indicate that the cost of deploying a BPL infrastructure is between $150 and $200 per home passed. Because only a portion of homes provided with a BPL capability will actually subscribe to the service, the actual cost per subscriber is far higher. For example, if 20 percent (or 1 out of every 5) of homes provided with a BPL service capability actually subscribe to the service, then the actual cost per subscriber would be between $750 ($150 × 5) and $1,000 ($200 × 5). Table 8.1 indicates the potential cost per subscriber at penetration rates ranging from 5 percent to 50 percent based on the cost per home passed averaging between $150 and $200.

In examining the entries in Table 8.1, note that a 5 percent penetration rate corresponds to 1 per 20 homes passed subscribing to a BPL service. Similarly, a 50 percent penetration rate corresponds to every other home passed actually subscribing to a BPL service. As the penetration rate increases, the cost per subscriber or, as more accurately shown in Table 8.1, the range of costs per subscriber decreases.

At a very low penetration rate of 10 percent, which corresponds to one home in ten that can use BPL actually subscribing, the cost per subscriber is between $1,500 and $2,000. If a utilities target rate of return is 10 percent, then it needs between $150 and $200 per year in income to justify the deployment of a BPL infrastructure. That income is after deducting maintenance and operational costs associated with a BPL deployment as well as billing expenses, allowance for deadbeats, and some type of Internet access fee splitting arrangement with an Internet Service Provider (ISP). Here the latter enables the utility

Table 8.1   BPL Average Cost per Subscriber

Penetration (%)	Cost per Subscriber (Range)
05	$3,000–4,000
10	1,500–2,000
15	1,000–1,333
20	750–1,000
25	600–800
30	500–666
35	428–571
40	375–500
45	333–444
50	300–400

backhaul to gain access to the Internet. Thus, the typical $34.95 per month fee charged during several field trials that resulted in a gross annual revenue of $419.40 may not be a sufficient ROI after maintenance, operation, and billing fees as well as ISP fee sharing are considered. Because SBC and Bell South recently lowered their DSL rates during the summer of 2005 to $19.95 and $24.95, respectively, utilities contemplating offering BPL may need to consider attempting to achieve a higher penetration rate by lowering their monthly fee below the rates charged for DSL if that technology is offered in their service area. Unfortunately, a price war used to attract cost-sensitive consumers could backfire if, for example, telephone companies and cable companies offering broadband service in the area of the electric utility also reduced their rates to match those of the utility. Thus, an introductory rate that offers a free modem and one or two months of free or reduced cost is probably more appealing than commencing a price war.

Technology

Four key technological issues currently act as impediments to the full-scale deployment of BPL: interoperability, radio interference, the use of WiFi versus the low-voltage (LV) power line, and the development of IP-addressable electric meters. In this section we will examine each of these technological impediments.

Interoperability

Although injectors, extractors, and repeaters have been used for numerous field trials, products manufactured by different vendors currently do not interoperate. Part of the problem concerning the lack of interoperability results from a lack of standards. Hopefully, the efforts of the HomePlug Powerline Alliance and the IEEE P1675 study group will alleviate the interoperability issue once standards are issued and vendors manufacture standard-compliant products.

Radio Interference

A second issue related to technology is radio interference caused by the use of orthogonal frequency division multiplexing (OFDM) equipment. Utilities need to consider applicable shielding as well as ensure compliance with Federal Communications Commission (FCC) regulations. Fortunately, the Electric Power Research Institute's report, discussed in Chapter 7, indicates how to minimize radio frequency risk factors associated with a BPL installation.

WiFi versus the LV Power Line

A third technological area that requires further study is the use of WiFi versus the LV power line to provide access from homes and offices onto the BPL infrastructure. The jury is still out on which method is better; however, the expected standardization of the IEEE 802.11n MIMO (multiple-in, multiple-out) technology may influence some utilities to consider WiFi, due to the extended transmission range afforded by the 802.11n standard over the 802.11b standard used in most WiFi field trials.

Automatic Electric Meter Capability

A fourth technological issue concerning BPL is its capability to provide an automatic electric meter reading. Currently there are several ways a meter reading capability can be accomplished. First, a utility can install a BPL modem on the customer's premises that uses an IP network address translation (NAT) capability to access an IP-address-able electric meter. Under this scenario, the customer would have to subscribe to the utility's BPL service and the utility would provide an IP-addressable electric meter and a NAT.

Because the use of IP-addressable electric meters would reduce the cost of meter reading, this application could provide a higher rate of return and increase the economic viability of BPL. Currently, reliable IP-addressable electric meters are in the prototype stage of development. Thus, this application is more than likely several years away from field trials. In addition, using a BPL modem on the subscriber's premises will require wiring to the meter, which is labor intensive and the cost of which could reduce any potential gains from automatically reading electric meters for a considerable period of time.

A second method a utility can consider to obtain an automated electric meter reading capability during a BPL rollout is to install a BPL modem and IP-addressable meter at each utility customer site that does not subscribe to the utility's BPL service. Although the co-location of the BPL modem and electric meter would reduce the cost of wiring the two together and eliminate the need for NAT, the utility would still face the cost of the modem and IP-addressable meter as well as the installation of the two devices. In addition, because the electric meter is normally mounted outdoors, the BPL modem would be required to be weatherproofed, further increasing the cost of the modem.

Similar to the argument concerning the use of IP-addressable electric meters installed at BPL subscriber premises, currently IP-addressable electric meters are in the development stage. Thus, this option is not presently available for utilities but could be a viable candidate for consideration within a few years.

A third option electric utilities can consider concerning addressable electric meters is a WiFi-based electric meter. Although this type of electric meter does not presently exist, it offers a mechanism to incorporate an IEEE 802.11n chip set into a solid-state electric meter that could be read at distances up to 300 feet from a utility power line. Thus, a rollout of BPL using a WiFi capability to service homes and small businesses would enable electric meters to be replaced on a gradual basis without requiring customers to subscribe to a BPL service. Because this potential method of eliminating the manual reading of meters also reduces personnel costs, it represents a mechanism for BPL providers to use the BPL infrastructure for minimizing other costs associated with operating a utility. In addition, for all automatic meter reading options discussed, a utility can conceivably offer new applications to their regular electric customers, such as time-of-day variances in the cost of electricity or using higher rates during peak consumption periods as a mechanism to discourage consumption that could require the utility to purchase power on the open market. Because such open market power purchases can be as high as $1,000 per megawatt, or five to ten times the normal wholesale cost of electricity, the ability to curb customer demand could have a significant effect on the bottom line of an electric utility.

Regulatory

Previously we noted that FCC regulations concerning emissions represent one example of the regulatory process that affects all types of transmission, including BPL, that radiate energy. In addition to FCC regulations concerning radiated energy and the list of frequencies that cannot be used by BPL modems, there are several potential regulatory areas that can affect both the deployment of BPL as well as the applications that can be offered by a BPL service. The regulatory issues we will discuss include taxes, quality of service (QoS) capability to provide Voice-over-IP (VoIP), and common carrier regulations.

Taxes

Regulators at both the federal and state levels may decide to treat BPL as another telecommunications business. If this occurs, all you need to do is look at your last telephone bill to determine the potential effect on BPL operators. In my geographic area, where telephone service is provided by Bell South, a $26 monthly bill includes approximately $10 of taxes and regulatory fees! If BPL service is regulated at the municipal level, then the possibility exists for additional fees. Needless to say, the additional costs in the form of taxes could easily damage the ability of utilities to achieve market penetration. On the other hand, if BPL is considered to represent an information service, it would not face many of the fees and taxes imposed on telecommunications services.

QoS and Common Carrier Issues

A second regulatory area that has not been addressed concerns applications that require a QoS capability, such as VoIP or Internet Telephony. First there is the question of how this service should be

Table 8.2   Advantages and Disadvantages of BPL

Advantages
Provides a new revenue source
Enables utilities to offer other applications
Utilizes utility infrastructure
BPL works

Disadvantages
Economic model uncertain
Interference an issue
Automatic electric meter reading not proven
Regulatory issues not resolved
Competition

regulated. If the BPL provider is considered to represent a common carrier, then it will be subject to further regulation by the FCC as well as by state regulatory commissions. This additional regulation can include the BPL service provider guaranteeing the routing of 911 calls and providing location tracking capability for calls. Thus, the common carrier issue could result in an increased expenditure to comply with additional regulations.

Now that we have an appreciation for the challenges facing BPL service providers, let's polish my crystal ball and peek into the future.

8.2 Predicting the Future

In attempting to predict the future of BPL, I must first review some of the advantages and disadvantages associated with deploying the service.

The key advantages and disadvantages associated with the deployment of BPL are summarized in Table 8.2. I will use the entries in the table as a basis for developing several predictions, rather than reviewing each entry.

Best Case Scenario

First, let us assume that each of the disadvantages listed in Table 8.2 can be overcome. That is, the economic model will indicate a viable rate of return due to a high penetration rate and interference as an issue will be favorably resolved; an economical solid-state electric meter with an IP addressing capability will be developed; regulatory issues will be resolved favorably for BPL service providers; and competition from cable and DSL will not undercut BPL service prices. Under this best case scenario we can probably expect a rapid deployment of BPL services over the next few years, resulting in approximately five to ten million homes and offices being served by 2010.

Worst Case Scenario

For a worst case prediction, let's assume that the disadvantages listed in Table 8.2 continue for the foreseeable future. Under this scenario, BPL will more than likely represent a limited service offered by municipalities that already have an infrastructure suitable for BPL, such as the City of Manassas (previously described in this book). In this situation, BPL can be expected to provide service to between 30 and 50 municipalities, with each municipality having between 2,000 and 5,000 potential subscribers. Thus, under this scenario, I believe the number of BPL subscribers could range between 60,000 and 250,000, or a very small fraction of broadband users.

Author's Prediction

In all likelihood, the deployment of BPL service will fall between the two previously mentioned scenarios. I believe that semirural areas that contain clusters of homes and offices will provide the best opportunity for the initial deployment of BPL, because such areas are more than likely not served by DSL nor cable modem providers. Later, I believe BPL providers could expand service to more rural areas and by 2010 have an installed base of customers ranging between 2.5 million and 5 million subscribers. Although this may appear low in comparison to the 20 million cable modem subscribers and 16 million DSL subscribers in the United States in 2005, from a revenue stream standpoint, it will significantly increase utility revenues. At a subscription fee of $30 per month, this would result in a gross revenue gain of $9 billion for utilities if the subscriber base expands to 2.5 million and $18 billion if the subscriber base expands to 5 million. Thus, a reasonable level of BPL penetration of the broadband market could provide electric utilities with a significant new source of revenue.

Index